THE GOLD Chemistry Experiments

How to Set Up a Home Laboratory— Over 200 Simple Experiments

BY ROBERT BRENT

ILLUSTRATED BY HARRY LAZARUS

Words Used by Chemists

Acid: a hydrogen-containing compound that releases hydrogen ions in solution.

Alloy: a material made up by combining two or more metals.

Analysis: breaking down a compound into two or more substances.

Anhydrous: free from water.

Atom: the smallest unit of an element that can enter into the making of a chemical compound.

Atomic weight: the weight of an atom compared with the weight of an oxygen atom set at 16.

Base: a compound containing the hydroxide group (OH).

Catalyst: a substance that helps in a chemical reaction without itself being changed.

Chemical change: a change of a substance into another substance having different properties.

Chemistry: a branch of science dealing with the compositions of substances and the changes that can be made in them.

Combustion: burning; a chemical change that produces heat and light.

Compound: a substance consisting of two or more different kinds of atoms in definite proportions by weight.

Crystal: a solid in which atoms or molecules are arranged in a definite pattern.

Density: the weight of a liquid or a solid in grams per cm^3 or milliliter.

Distillate: a liquid that has been turned into vapor and again cooled into a liquid.

Distillation: the process of producing a distillate.

Ductile: capable of being drawn out into a wire.

Electrolysis: breaking down a substance by passing an electric current through it.

Electrolyte: a substance that, when in a solution or when melted, will conduct an electric current.

Element: a substance that contains only one kind of atoms.

Equation: a complete description of a chemical reaction by the use of symbols, formulas, and signs.

Evaporation: the changing of a substance into vapor; also the process of removing water by heating.

Filtrate: a liquid obtained by filtration.

Filtration: the process of straining a liquid from a solid through porous material, usually filter paper.

Formula: a group of symbols and numbers giving the composition of a compound.

Hydrate: a compound containing loosely bound water of hydration (water of crystallization) that can be driven off by heating.

Hydroxide: a compound that contains the hydroxyl (OH) radical.

Ion: an electrically charged atom or group of atoms (radical).

Malleable: capable of being hammered or rolled into a thin sheet.

Matter: anything that takes up space and has weight.

Metal: an element that is a good conductor of electricity, has luster, and whose oxide forms a base with water.

Metalloid: an element that has properties of both metals and nonmetals.

Mixture: a mingling of substances not combined chemically.

Molecular weight: the sum of the atomic weights of the atoms that make up a molecule of a compound.

Molecule: the smallest unit of a compound that can exist in the free state.

Neutralization: the reaction of an acid and a base to give a salt and water.

Nonmetal: an element that is a poor conductor of electricity, does not have luster, and whose oxide forms an acid when combined with water.

Organic chemistry: the chemistry of the carbon compounds.

Oxidation: the process by which a substance combines with oxygen.

Precipitate: an insoluble solid formed in a solution by chemical reaction.

Radical: a group of atoms that behave chemically as a single atom.

Reaction: a chemical change.

Reduction: removal of oxygen; the opposite of oxidation.

Salt: compound (other than water) formed by the reaction of an acid and a base.

Saturated solution: a solution that contains the maximum amount of solute under the conditions.

Solubility: the number of grams of a solute needed to make a saturated solution in 100 grams of solvent.

Solute: the substance dissolved in a solvent.

Solution: a non-settling mixture of a solute in a solvent.

Solvent: a liquid in which a solute is dissolved.

Sublimation: a process by which a solid is turned into vapor and again cooled into a solid without passing through a liquid stage.

Subscript: a small numeral indicating the number of atoms of a certain element in the formula of a compound.

Substance: any specific kind of matter whether element, compound, or mixture.

Symbol: a letter or two letters representing one atom of an element.

Synthesis: the making up of a compound from simpler compounds or from elements; the opposite of analysis.

Valence: the number of hydrogen atoms which one atom of an element can displace or with which it can unite.

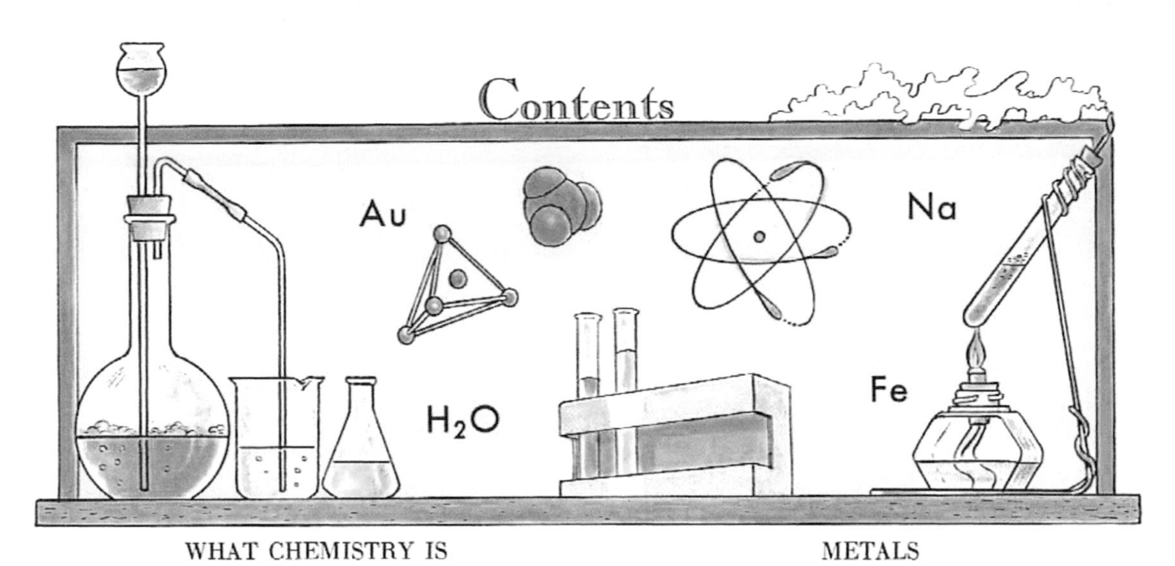
Contents
Au
Na
H_2O
Fe

EVERY HOME KITCHEN IS A CHEMICAL LABORATORY.
COOKING AND CLEANING ARE CHEMICAL PROCESSES.

The Importance of Chemistry

THERE IS HARDLY a boy or a girl alive who is not keenly interested in finding out about things. And that's exactly what chemistry is: FINDING OUT ABOUT THINGS — finding out what things are made of and what changes they undergo.

What things? Any thing! Every thing!

Take a look around you. All the things you see — and lots of things you can't see — have to do with the science of chemistry.

Let's start with yourself. The air you breathe is a mixture of chemical substances and the process of breathing is a chemical reaction. The foods you eat are all chemical products and the ways in which your body turns them into muscles and bones and nerves and brain cells are some of the greatest of all chemical mysteries.

The clothes you wear, the books you read, the medicine you take, the house in which you live — all these are products of chemistry. So is the family car — the metal in it, the rubber on which it rolls, the gas that moves it.

Nature itself is a tremendous chemical laboratory. Everything in nature is forever passing through chemical changes. Here on earth, plants and animals

INSECT SPRAYS MEAN HEALTHIER LIVESTOCK.

CHEMISTRY PROVIDES FUEL FOR ALL KINDS OF TRANSPORTATION.

PURIFICATION OF WATER

grow, die, and decay; rocks crack and crumble under the influence of air and water. In the universe, new stars are formed, others fade. The sun that gives us heat and light and energy is a flaming furnace of chemical processes that will eventually burn itself out, billions of years from now.

Chemistry is one of the most important of all sciences for human welfare.

Chemistry means the difference between poverty and starvation and the abundant life. The proper use of chemistry makes it possible for farmers to feed the world's ever-increasing population, for engineers to develop new means of transportation and communication that will bring the peoples of the world closer together, for doctors to cure the diseases of mankind, for manufacturers to produce the thousands of items that are necessary for better and richer living.

And this is only the beginning.

Within recent years, scientists have succeeded in penetrating into the innermost secrets of chemical substances and have begun to make use of the tremendous force that lies hidden in them. This atomic power opens up amazing possibilities for the future.

You will live in a world in which chemistry will become ever more important. To understand that world it is necessary to understand the truths and laws on which modern chemistry is based and to learn how chemists of the past unraveled them.

This book will help you get this insight — not alone by your reading it, but also by your conscientiously doing the experiments described and learning what each of them has to tell you.

OIL IS THE BASIS FOR COUNTLESS CHEMICAL PRODUCTS.

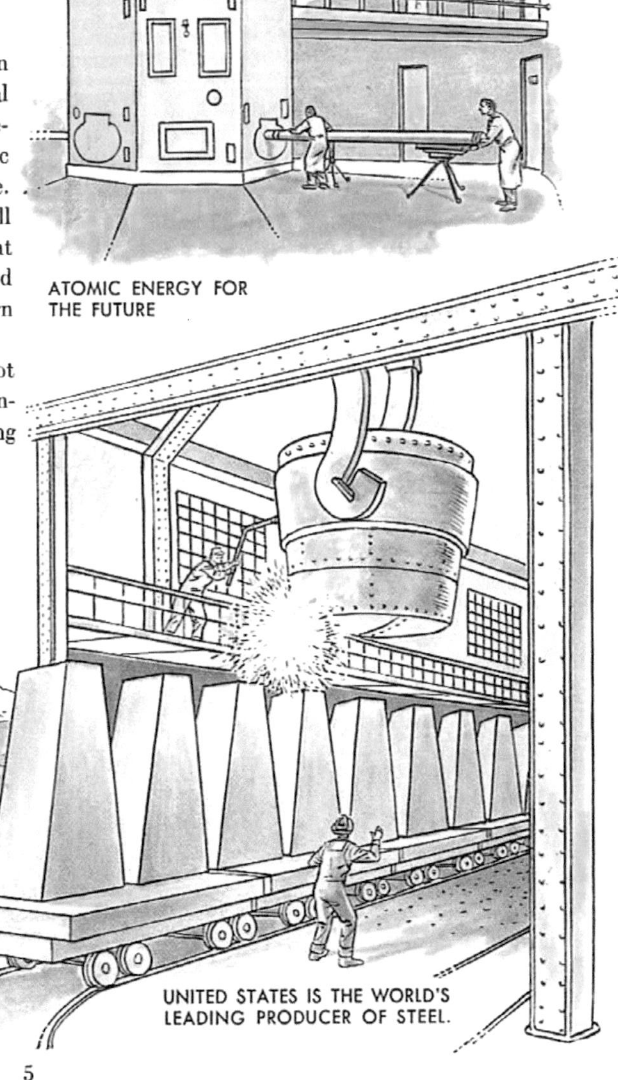

ATOMIC ENERGY FOR THE FUTURE

UNITED STATES IS THE WORLD'S LEADING PRODUCER OF STEEL.

PAPER AND PRINTERS' INK ARE MADE WITH HELP OF CHEMISTRY.

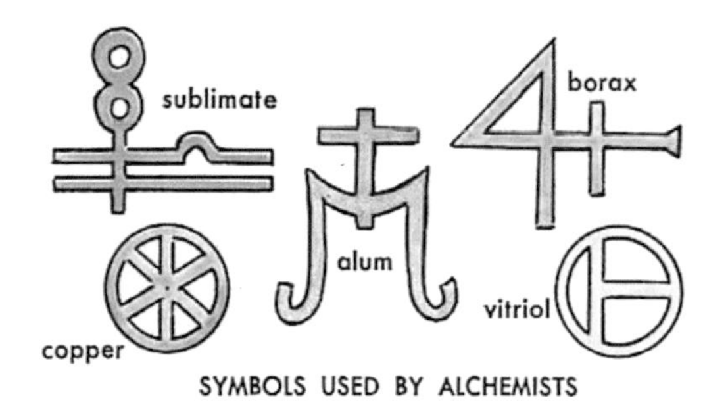

SYMBOLS USED BY ALCHEMISTS

Chemists of the Past

MANY THOUSAND years ago, an early ancestor of yours pushed a stick into the hot lava flowing from an erupting volcano. The stick burst into fire. He held it up as a torch. It gave off light and heat and finally turned into ashes.

This ancient man might be considered the world's first chemist. He had actually taken a substance called wood and had, by a chemical process called combustion or burning, turned it into something else.

The discovery of the use of fire was the first great step leading toward modern chemistry. Fire made it possible to turn raw foodstuffs into edible meals, to bake shaped clay into pottery, to make glass, to drive metals out of their ores.

For thousands of years people were chiefly interested in the results of what they did — they didn't care about what happened or why it happened. It was only about 2,500 years ago that philosophers began to wonder about what things were made of and what happened when a thing changed into something else.

Around 400 B. C., in Greece, a thinker by the name of Empedocles came up with an idea that seemed to make sense. He explained that everything in the world was made from just four things which he called "elements": fire, water, air, and earth. Think of that burning stick mentioned above. It gave off fire — so, obviously, the stick had to contain fire. It sizzled — which meant there was water in it. It smoked — and smoke would be some kind of air. It left ashes — and ashes are earth, as certainly everyone should know.

Everyone — except another Greek, Democritus, born around the time when Empedocles died. He had a different notion — that all matter was made up of tiny particles which he called *atomos* — something that cannot be cut further.

But Democritus didn't get very far with his idea. The greatest Greek philosopher of the day, Aristotle, held out for the four elements. And because of his great reputation this false idea governed the thinking of scientists for two thousand years — because no one dared suggest that he knew better than the great Aristotle!

BRONZE-AGE MAN WAS ONE OF THE EARLIEST CHEMISTS.

DEMOCRITUS INSISTED THAT MATTER CONSISTS OF ATOMS.

PARACELSUS TOLD HIS PUPILS TO USE EXPERIMENTS.

BOYLE INVESTIGATED GASES AND BROKE OLD TRADITIONS.

In the meantime, scientists of Arabia began work in a subject they called alchemy — from Arabic *al*, the, and *kimia*, pouring together. They mixed things and boiled and distilled and extracted in the hope, some day, of finding a way of making GOLD! They discovered a great number of things not previously known, developed many sound laboratory methods, and gave the science of chemistry its name — but they never created the slightest speck of gold. Neither did a great number of European alchemists.

For hundreds of years chemistry made little headway. Then, in 1525, a Swiss doctor and scientist spoke up. He had the imposing name of Theophrastus Bombastus Paracelsus von Hohenheim. He challenged his students to tear up their books with the old theories that had been developed through reasoning only and to find out for themselves *through experiments* whether a scientific theory was right or wrong. But only a few people paid attention to him.

More than a hundred years passed before an Englishman, Robert Boyle, in 1661, succeeded in killing off the old idea of the four elements. He did it by establishing that there are many elements — substances that cannot be formed by other substances and cannot be broken into other substances.

Another hundred years went by. Then, at the time of the American Revolution, the day finally dawned for modern chemistry.

A Swede, Karl Scheele, and an Englishman, Joseph Priestley, discovered oxygen, and a Frenchman, Antoine Laurent Lavoisier, explained the true nature of burning and made up the first scientific listing of all known elements — twenty-eight at the time.

Within a few years, more elements were found. With the help of electricity, an English chemist, Humphry Davy, in a single year brought to light six new metals — among them sodium, potassium, calcium, and magnesium.

Twenty years later, in 1828, another important break-through occurred. A German chemist, Friedrich Wohler, working in his laboratory produced a chemical, urea, that had never before been made outside the body of a living animal.

More and more things were happening. New elements were discovered, new chemicals created. The advances in chemistry greatly influenced industry, agriculture and medicine.

And then, in 1898, the Polish-born Marie Curie and her French husband, Pierre, discovered the "miracle element," radium. This opened up a whole new age in chemistry.

Within the last fifty years, chemistry has moved forward with giant steps. But not a single one of these steps would have been possible without the dedicated work of the chemists of the past who laid the foundation on which modern chemistry rests.

PRIESTLEY USED HEAT OF SUN TO PRODUCE OXYGEN.

LAVOISIER GAVE THE RIGHT EXPLANATION OF BURNING.

DAVY BROUGHT ELECTRICITY INTO CHEMICAL RESEARCH.

MARIE CURIE AND HER HUSBAND DISCOVERED RADIUM.

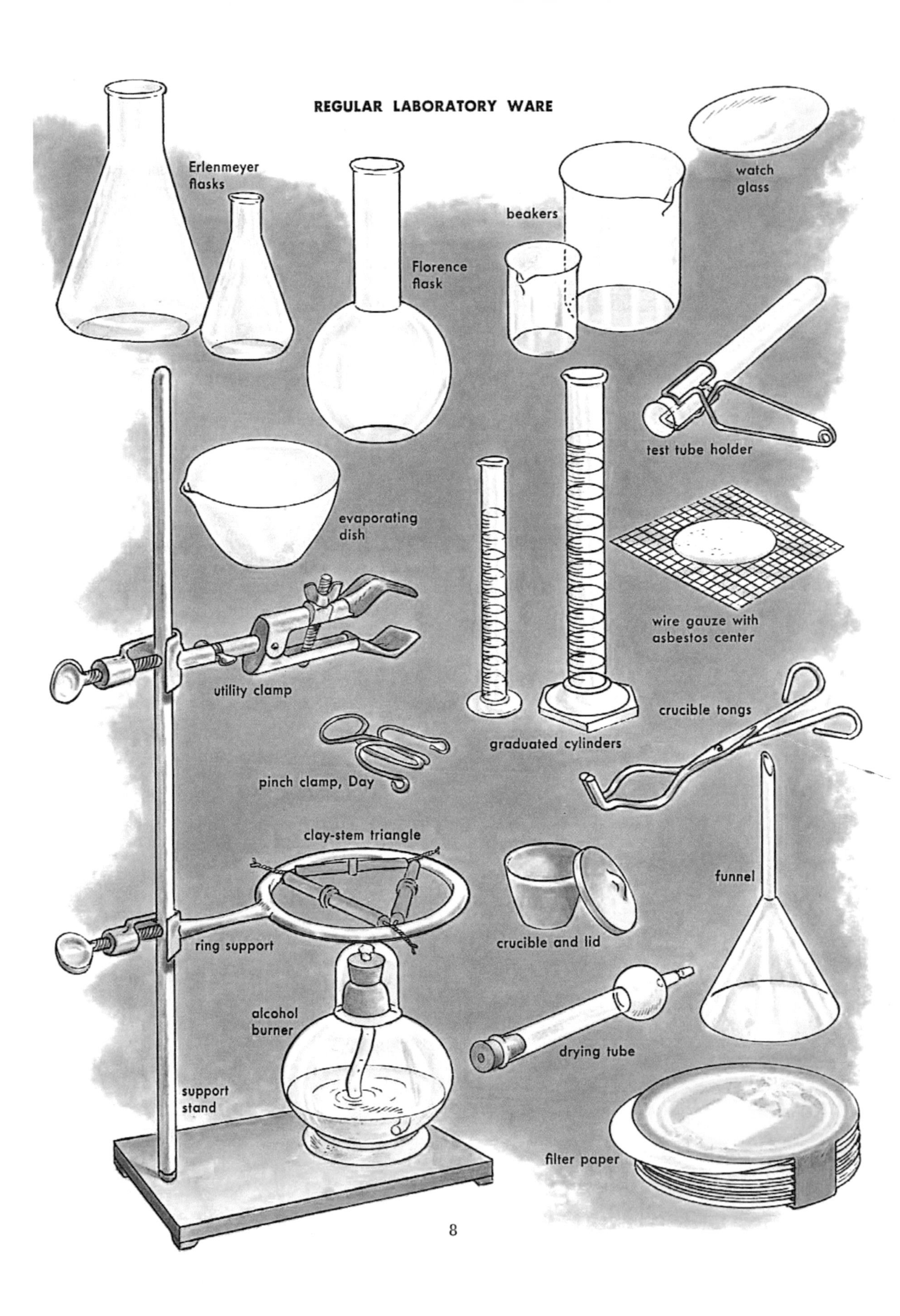
REGULAR LABORATORY WARE
Erlenmeyer flasks
Florence flask
beakers
watch glass
test tube holder
evaporating dish
wire gauze with asbestos center
utility clamp
graduated cylinders
crucible tongs
pinch clamp, Day
clay-stem triangle
funnel
ring support
crucible and lid
alcohol burner
drying tube
support stand
filter paper

Equipment for Chemistry

SOME of the greatest discoveries in chemistry were made by scientists who had no special equipment but simply used whatever was at hand.

In your home lab experiments it will pay you to follow the example of these early chemists. Put your imagination to work. Use whatever suitable equipment you can find around the house (as suggested in column to the right) and buy only what is absolutely necessary (as shown below). Some items may be purchased in a local drugstore or scientific supply shop. If not, you can buy them from one of the suppliers listed on page 110.

Later on — if you really get excited about chemistry — you may want to use your pocket money for some of the lab equipment shown on page 8.

LABORATORY WARE FOR HOME LAB

test tubes, Pyrex
16 mm x 150 mm

wide-mouth
bottle, 4 oz.

rubber stoppers
numbers 00 to 6

red and
blue
litmus
paper

triangular file

medicine
dropper

glass tubing
5 mm outside diameter

glass rod

test tube
brush

rubber tubing
$\frac{3}{16}$" inside diameter

IMPROVISED EQUIPMENT FOR HOME LAB

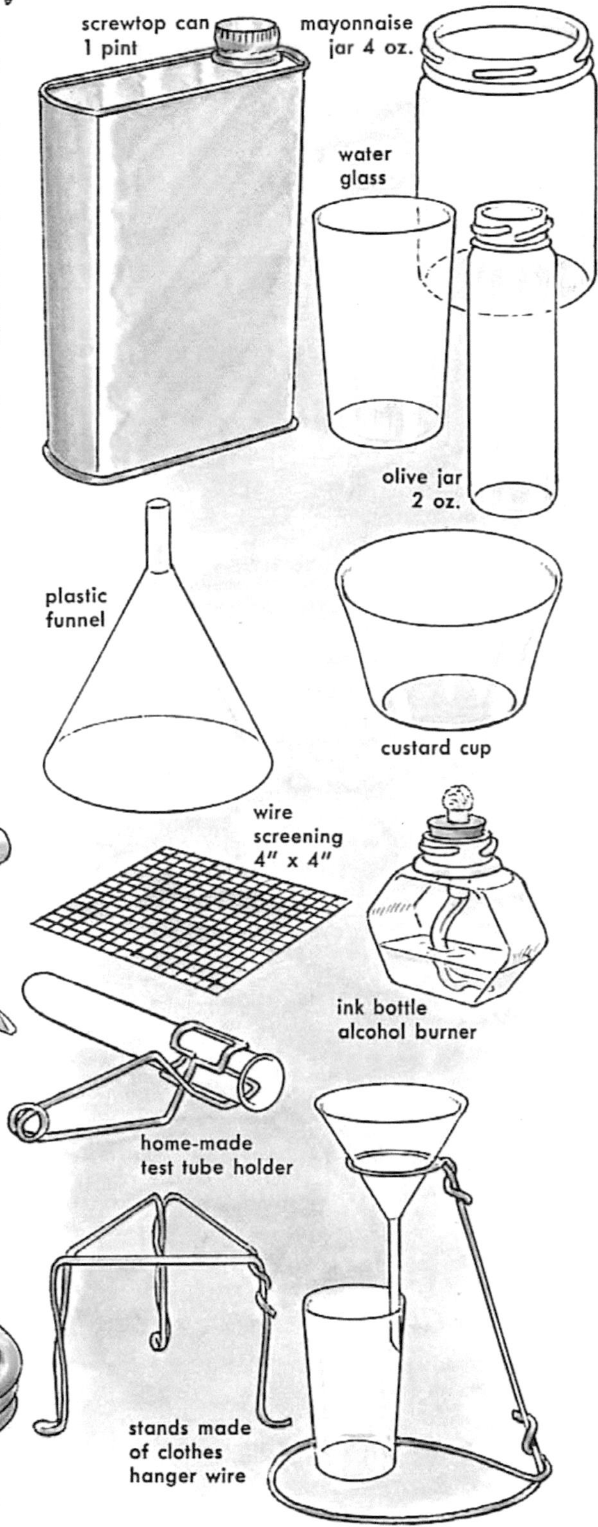

Setting Up Your Home Laboratory

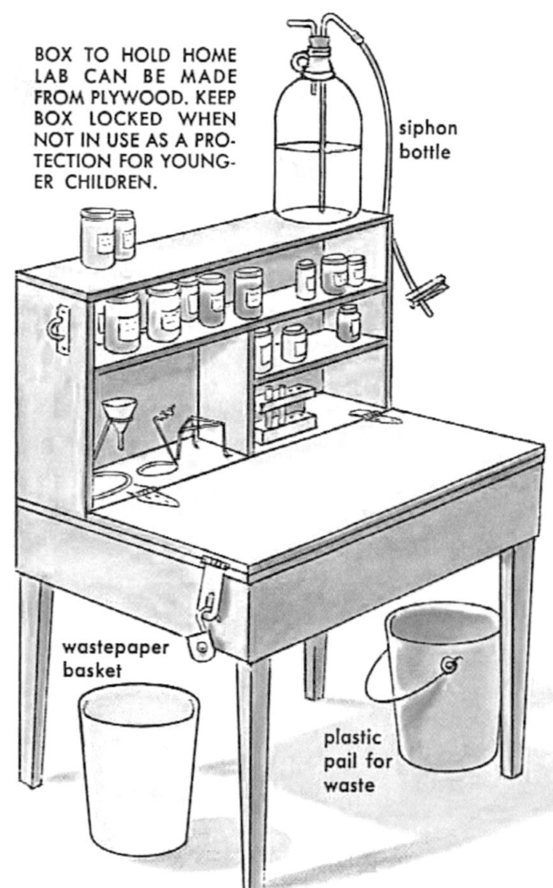

It is possible that you may be permitted to work at the kitchen table when this is not in use. But it is far better if you have a place where you will not be disturbed and where you can store your equipment — a corner in your room, or in the basement or the garage.

These are the things you'll need in your lab:

Work Table. An old, sturdy table will do. Cover it with a plastic top to protect the wood.

Water Supply. If you have a faucet nearby, fine. Otherwise, make a siphon bottle (page 11).

Waste Disposal. If you can dump your waste directly into the kitchen *drain* (NOT into the sink), you are all right. If not, collect it in a plastic pail to be thrown out when you're finished.

Source of Heat. In the regular laboratory, special gas burners are used. In the home lab, you can use a burner for denatured alcohol. Have a shallow metal pan under the burner for fire safety.

Storage. If there's no one around to disturb your chemicals and equipment, an open shelf is OK. Otherwise, use a box that can be locked up.

Containers. Keep chemicals in glass jars and bottles. LABEL THEM ALL CLEARLY.

Stands. Make your own test tube stand as well as stands for holding glassware for heating.

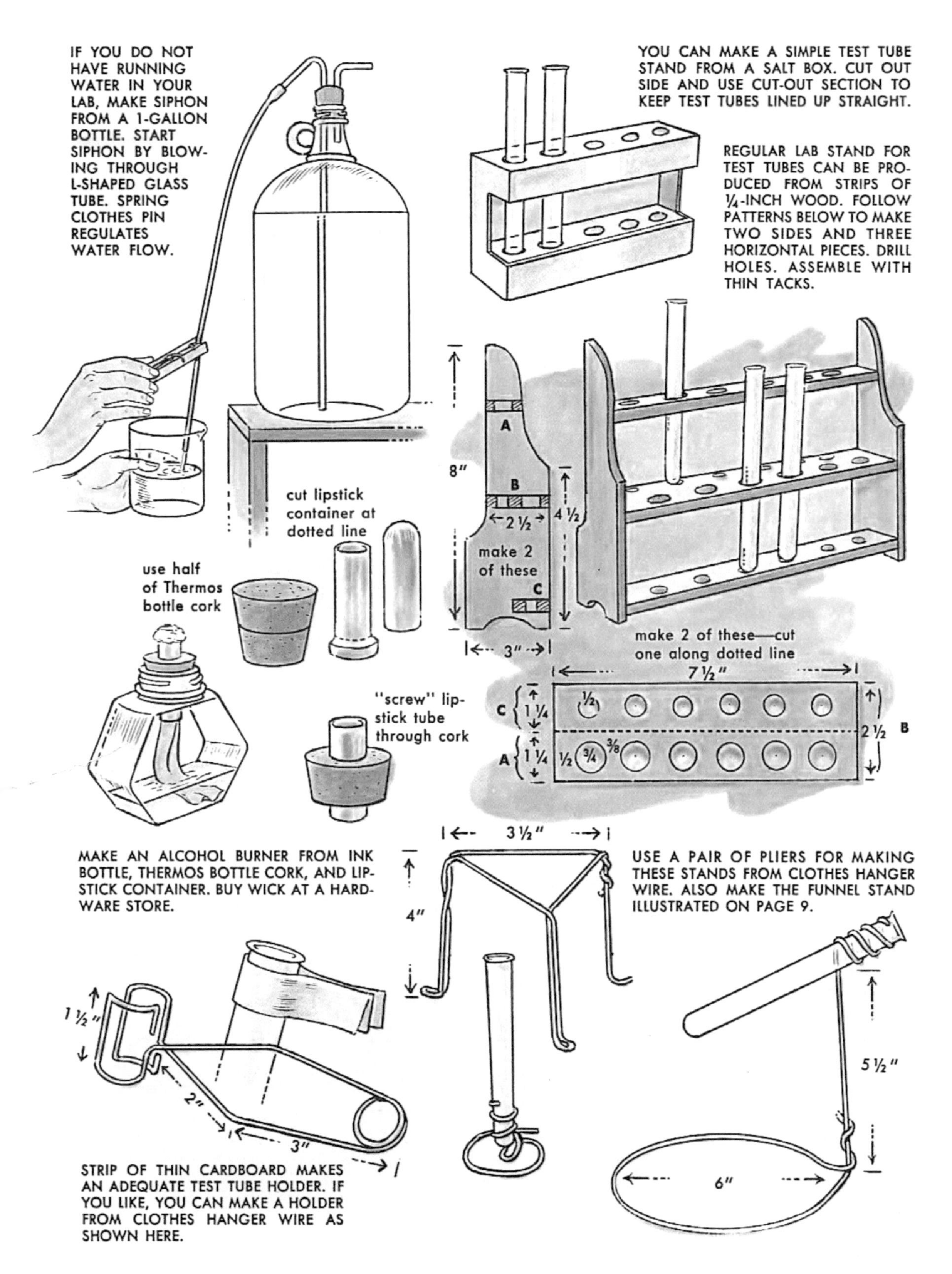

IF YOU DO NOT HAVE RUNNING WATER IN YOUR LAB, MAKE SIPHON FROM A 1-GALLON BOTTLE. START SIPHON BY BLOWING THROUGH L-SHAPED GLASS TUBE. SPRING CLOTHES PIN REGULATES WATER FLOW.

YOU CAN MAKE A SIMPLE TEST TUBE STAND FROM A SALT BOX. CUT OUT SIDE AND USE CUT-OUT SECTION TO KEEP TEST TUBES LINED UP STRAIGHT.

REGULAR LAB STAND FOR TEST TUBES CAN BE PRODUCED FROM STRIPS OF 1/4-INCH WOOD. FOLLOW PATTERNS BELOW TO MAKE TWO SIDES AND THREE HORIZONTAL PIECES. DRILL HOLES. ASSEMBLE WITH THIN TACKS.

MAKE AN ALCOHOL BURNER FROM INK BOTTLE, THERMOS BOTTLE CORK, AND LIPSTICK CONTAINER. BUY WICK AT A HARDWARE STORE.

USE A PAIR OF PLIERS FOR MAKING THESE STANDS FROM CLOTHES HANGER WIRE. ALSO MAKE THE FUNNEL STAND ILLUSTRATED ON PAGE 9.

STRIP OF THIN CARDBOARD MAKES AN ADEQUATE TEST TUBE HOLDER. IF YOU LIKE, YOU CAN MAKE A HOLDER FROM CLOTHES HANGER WIRE AS SHOWN HERE.

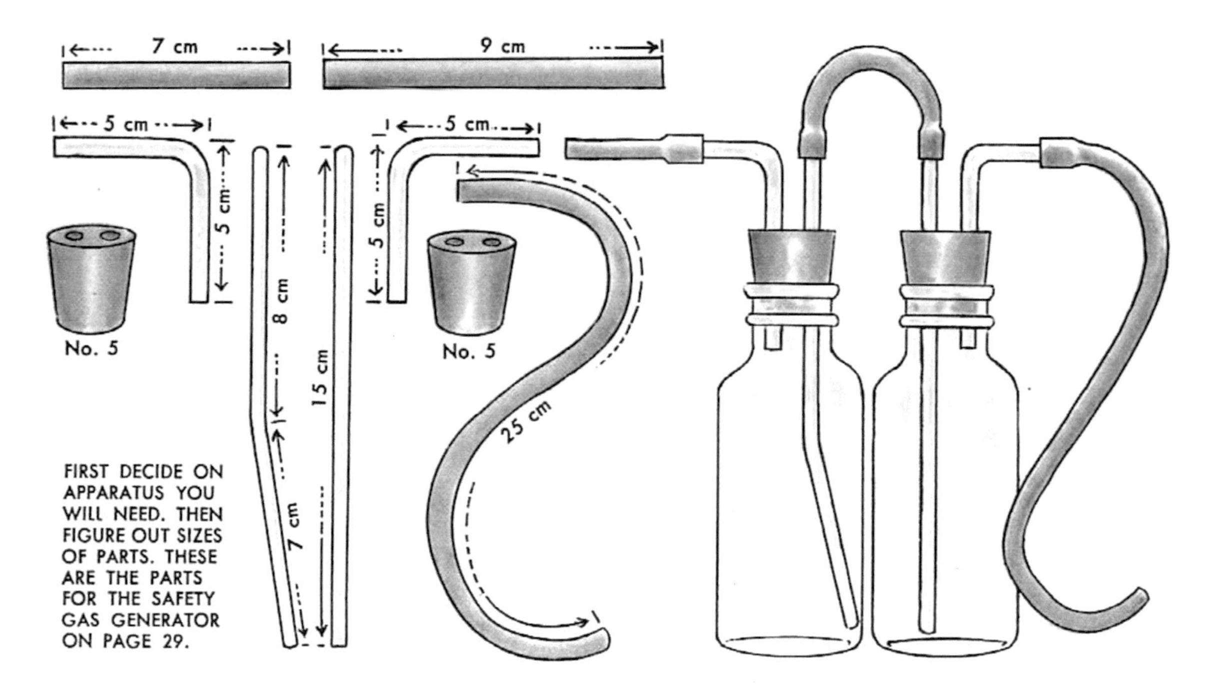

Making Apparatus for Experiments

Most of your chemical experiments you will perform in test tubes and jars. But occasionally you will need an apparatus — a device consisting of bottles and stoppers, glass and rubber tubing.

A good chemist takes pride in his apparatus. He makes it with great care — not just for looks but, more important, for safety. An apparatus that leaks flammable gas can be very dangerous.

Before you start to put an apparatus together, make a simple drawing of it so that you will know what it will consist of. Then get out the various parts you will need to put it together.

To make an apparatus, you need to know how to cut a glass tube, how to bend it, and how to draw it to a jet point. See page 13.

It is wise to use glass tubes of one diameter only, with rubber tubing to fit. Glass tubes of an *outside* diameter of 6 millimeters fit snugly into the holes in the usual rubber stoppers. Rubber tubing of an *inside* diameter of 3/16″ fits over the 6mm glass tubes.

To determine the right size stoppers to use in the bottles of your apparatus, measure the mouths of the bottles against the stoppers shown below in actual size. Order stoppers by number. Keep a selection of different sizes on hand.

Follow the safety precautions on page 16.

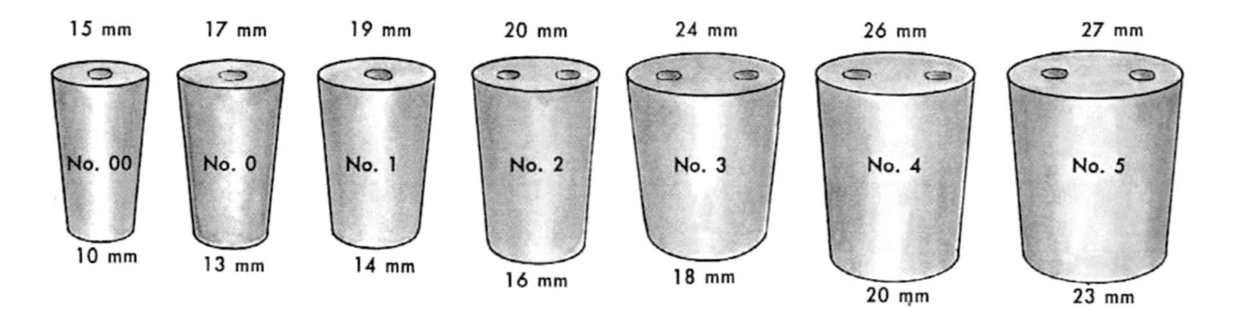

THESE ARE THE ACTUAL SIZES OF RUBBER STOPPERS. BY MEASURING THEM AGAINST YOUR LAB WARE YOU WILL KNOW WHICH TO ORDER. No. 0 FITS THE 16 mm TEST TUBE. No. 5 FITS 4-OZ. WIDE-MOUTH BOTTLE.

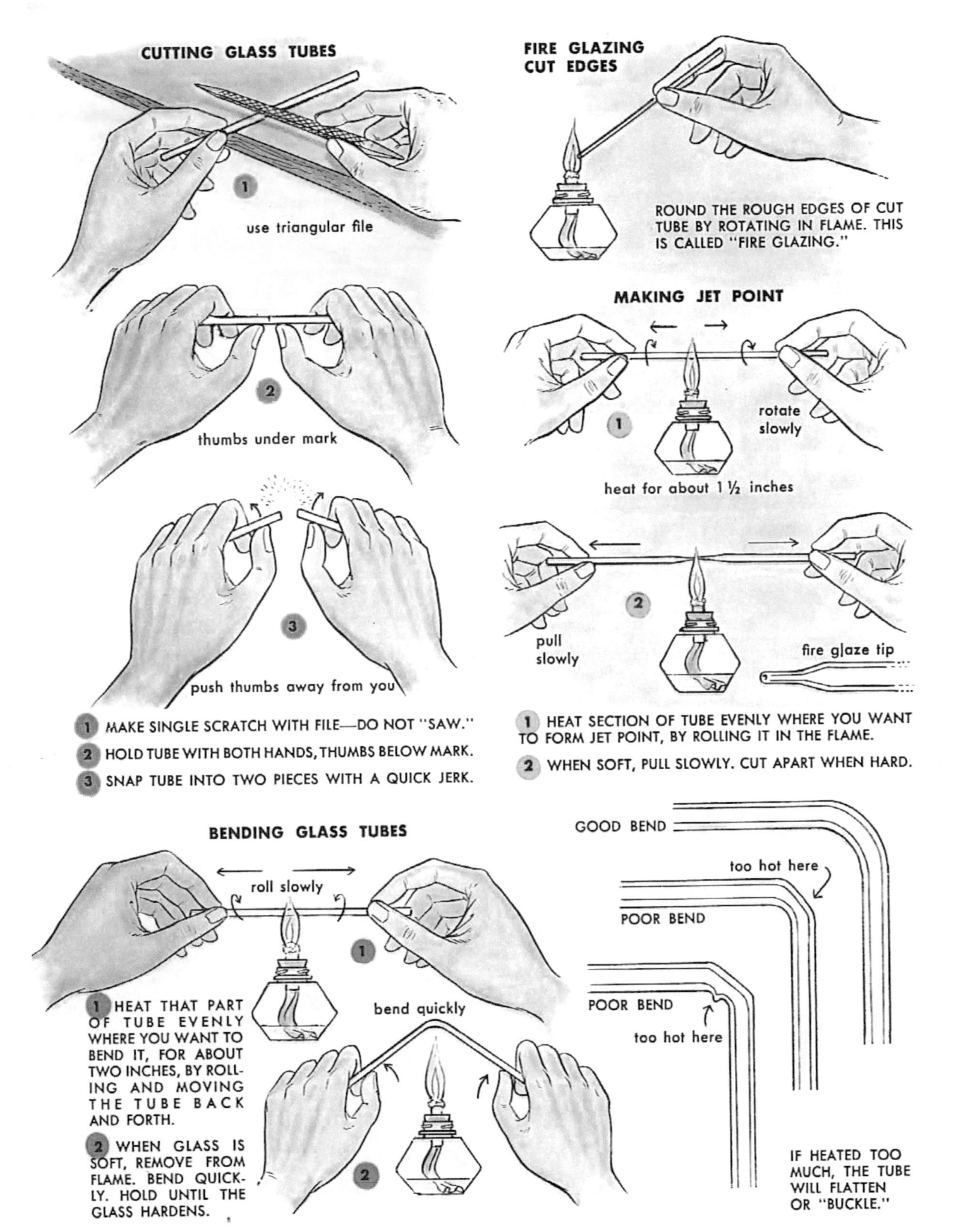
CUTTING GLASS TUBES
1
use triangular file
2
thumbs under mark
3
push thumbs away from you
1 MAKE SINGLE SCRATCH WITH FILE—DO NOT "SAW."
2 HOLD TUBE WITH BOTH HANDS, THUMBS BELOW MARK.
3 SNAP TUBE INTO TWO PIECES WITH A QUICK JERK.
FIRE GLAZING CUT EDGES
ROUND THE ROUGH EDGES OF CUT TUBE BY ROTATING IN FLAME. THIS IS CALLED "FIRE GLAZING."
MAKING JET POINT
1
rotate slowly
heat for about 1 ½ inches
2
pull slowly
fire glaze tip
1 HEAT SECTION OF TUBE EVENLY WHERE YOU WANT TO FORM JET POINT, BY ROLLING IT IN THE FLAME.
2 WHEN SOFT, PULL SLOWLY. CUT APART WHEN HARD.
BENDING GLASS TUBES
roll slowly
1
1 HEAT THAT PART OF TUBE EVENLY WHERE YOU WANT TO BEND IT, FOR ABOUT TWO INCHES, BY ROLLING AND MOVING THE TUBE BACK AND FORTH.
bend quickly
2 WHEN GLASS IS SOFT, REMOVE FROM FLAME. BEND QUICKLY. HOLD UNTIL THE GLASS HARDENS.
2
GOOD BEND
too hot here
POOR BEND
POOR BEND
too hot here
IF HEATED TOO MUCH, THE TUBE WILL FLATTEN OR "BUCKLE."

Scientific Measurements

In science, the metric system is preferred over our usual system. It is much easier to work with when once you have learned it — for instead of dividing or multiplying by 12 or 32 or 16 to go from one unit to the next, you simply move the decimal point. Just remember these two things:

1. That the names of the basic units are meter for lengths, liter for volumes, grams for weights — abbreviated to m, l, and g (without a period after them).
2. That 1000 of a kind are called kilo; 100, hekto; 10, deca; 1/10 is called deci; 1/100, centi; 1/1000, milli.

METRIC UNITS OF LENGTH

1000 meters (m) = 1 kilometer (km)
1 meter (m) = 1000 millimeters (mm)
1 meter (m) = 39.37 inches
2.540 centimeters (cm) = 1 inch

METRIC UNITS OF VOLUME

1 liter (l) = 1000 cubic centimeters (cm^3 or cc)
1 liter (l) = 1000 milliliters (ml)
1 liter (l) = 1.06 quarts (liquid)
0.946 liter (l) = 1 quart (liquid)

METRIC UNITS OF WEIGHT

1000 grams (g) = 1 kilogram (kg)
1 gram (g) = 1000 milligrams (mg)
1 gram (g) = 0.035 ounces avoirdupois
28.350 grams (g) = 1 ounce avoirdupois

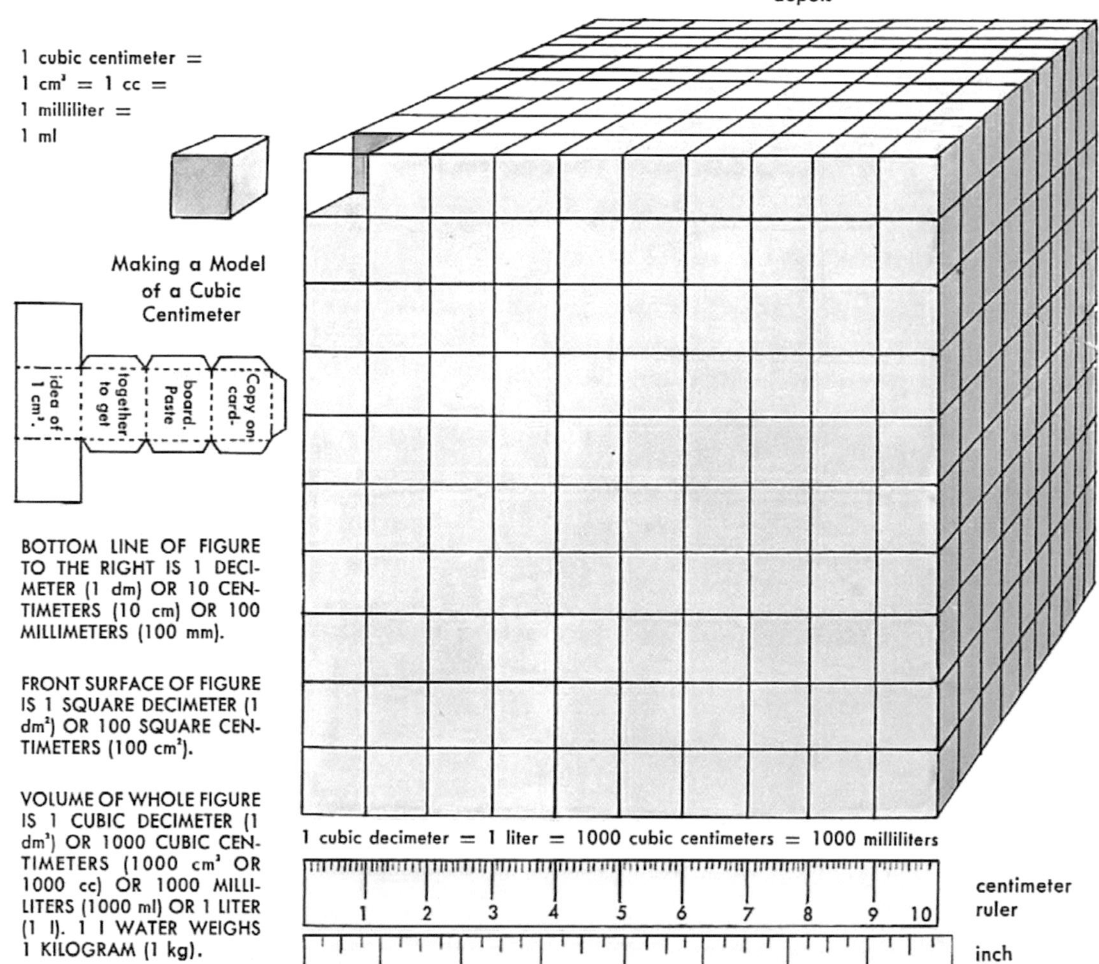

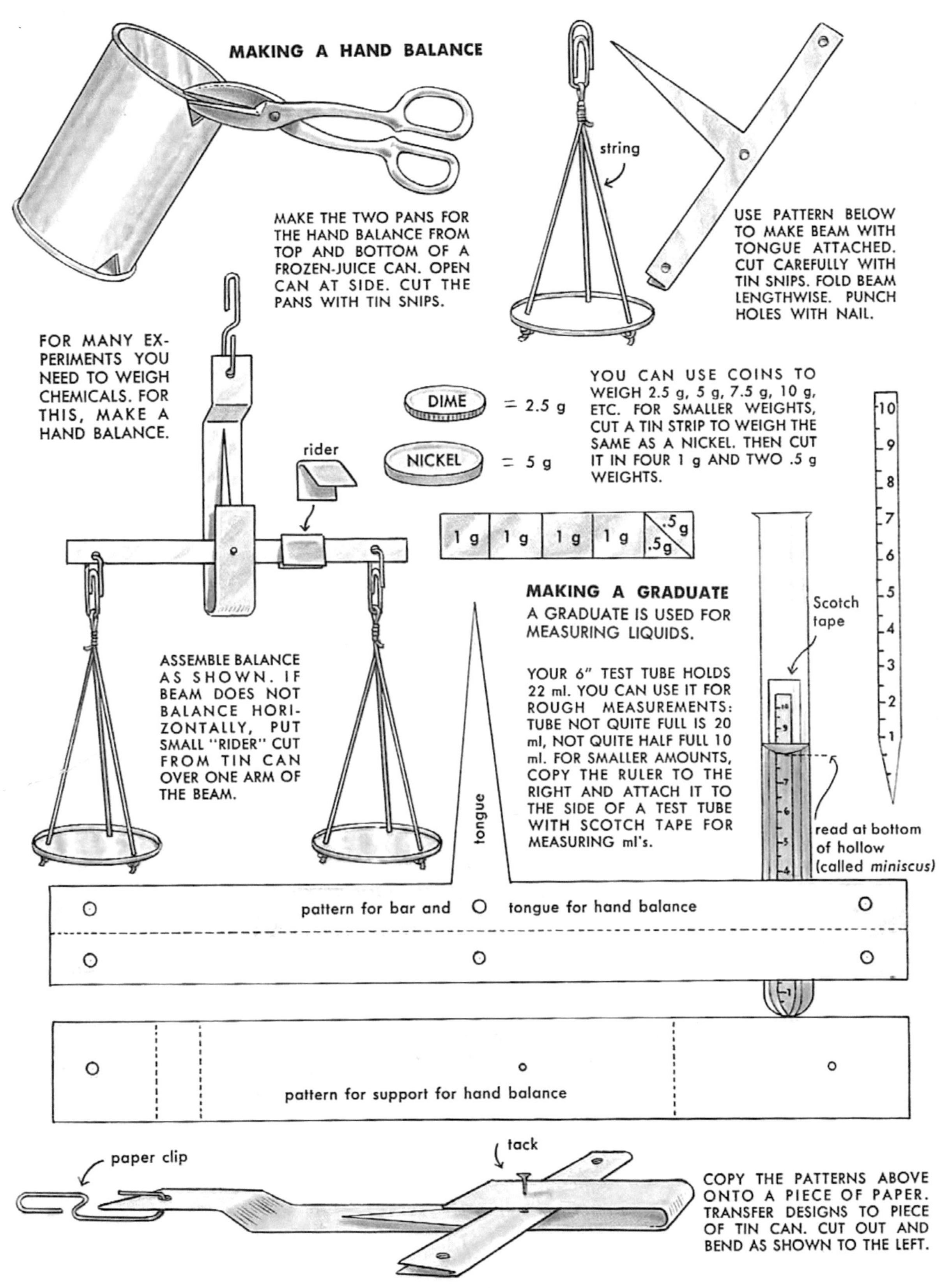
MAKING A HAND BALANCE
MAKE THE TWO PANS FOR THE HAND BALANCE FROM TOP AND BOTTOM OF A FROZEN-JUICE CAN. OPEN CAN AT SIDE. CUT THE PANS WITH TIN SNIPS.
string
USE PATTERN BELOW TO MAKE BEAM WITH TONGUE ATTACHED. CUT CAREFULLY WITH TIN SNIPS. FOLD BEAM LENGTHWISE. PUNCH HOLES WITH NAIL.
FOR MANY EXPERIMENTS YOU NEED TO WEIGH CHEMICALS. FOR THIS, MAKE A HAND BALANCE.
DIME = 2.5 g
NICKEL = 5 g
YOU CAN USE COINS TO WEIGH 2.5 g, 5 g, 7.5 g, 10 g, ETC. FOR SMALLER WEIGHTS, CUT A TIN STRIP TO WEIGH THE SAME AS A NICKEL. THEN CUT IT IN FOUR 1 g AND TWO .5 g WEIGHTS.
rider
1 g
1 g
1 g
1 g
.5 g
.5 g
MAKING A GRADUATE
A GRADUATE IS USED FOR MEASURING LIQUIDS.
YOUR 6" TEST TUBE HOLDS 22 ml. YOU CAN USE IT FOR ROUGH MEASUREMENTS: TUBE NOT QUITE FULL IS 20 ml, NOT QUITE HALF FULL 10 ml. FOR SMALLER AMOUNTS, COPY THE RULER TO THE RIGHT AND ATTACH IT TO THE SIDE OF A TEST TUBE WITH SCOTCH TAPE FOR MEASURING ml's.
Scotch tape
read at bottom of hollow (called miniscus)
10
9
8
7
6
5
4
3
2
1
ASSEMBLE BALANCE AS SHOWN. IF BEAM DOES NOT BALANCE HORIZONTALLY, PUT SMALL "RIDER" CUT FROM TIN CAN OVER ONE ARM OF THE BEAM.
tongue
pattern for bar and tongue for hand balance
pattern for support for hand balance
paper clip
tack
COPY THE PATTERNS ABOVE ONTO A PIECE OF PAPER. TRANSFER DESIGNS TO PIECE OF TIN CAN. CUT OUT AND BEND AS SHOWN TO THE LEFT.

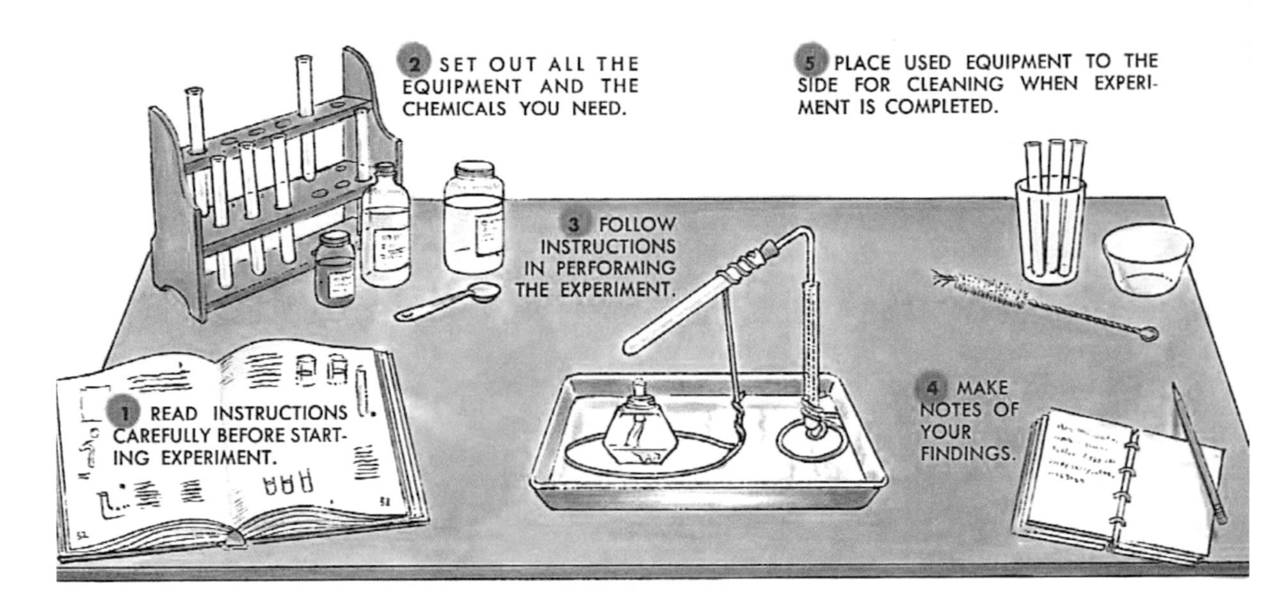

Correct Laboratory Techniques

In your home laboratory, three considerations are of the greatest importance: SAFETY, NEATNESS, and EXACTNESS.

SAFETY — All the experiments in this book are safe when done in the correct laboratory way as shown on these pages.

Treat chemicals with respect. Never taste anything unless specifically told to do so. If there are younger children in the family, lock up your chemicals when you are not working with them.

Protect your clothes with a plastic apron.

Be careful with fire. When you use your alcohol burner, have a metal pan under it for safety.

NEATNESS — Get the habit of lining up equipment and chemicals you need on one side and placing used items on the opposite side — keeping the space between them clear for your experiments.

Put chemicals away and clean glassware as soon as you have finished an experiment.

EXACTNESS — Label all bottles and jars containing chemicals clearly and correctly.

Where amounts of chemicals are not given, use the smallest amount that will tell you what you want to know.

Observe the chemical reactions carefully and make complete notes of them as you go along.

PLAY SAFE WHEN YOU PUT A GLASS TUBE IN A STOPPER. PROTECT YOUR HANDS BY WRAPPING TOWEL AROUND THEM. MOISTEN GLASS TUBE AND STOPPER WITH WATER, THEN PUSH THE TUBE INTO THE STOPPER WITH A SCREW-DRIVER MOTION.

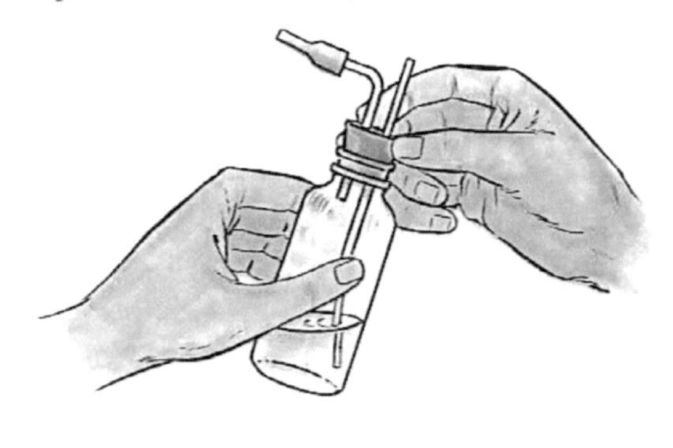

WHEN YOU MAKE AN APPARATUS FOR A CHEMICAL EXPERIMENT, MAKE SURE THAT ALL CONNECTIONS ARE AIR-TIGHT. USE THE RIGHT SIZE STOPPER FOR MOUTH OF CONTAINER, GLASS TUBES THAT FIT SNUGLY INTO THE STOPPER HOLES, AND TIGHT-FITTING RUBBER TUBING.

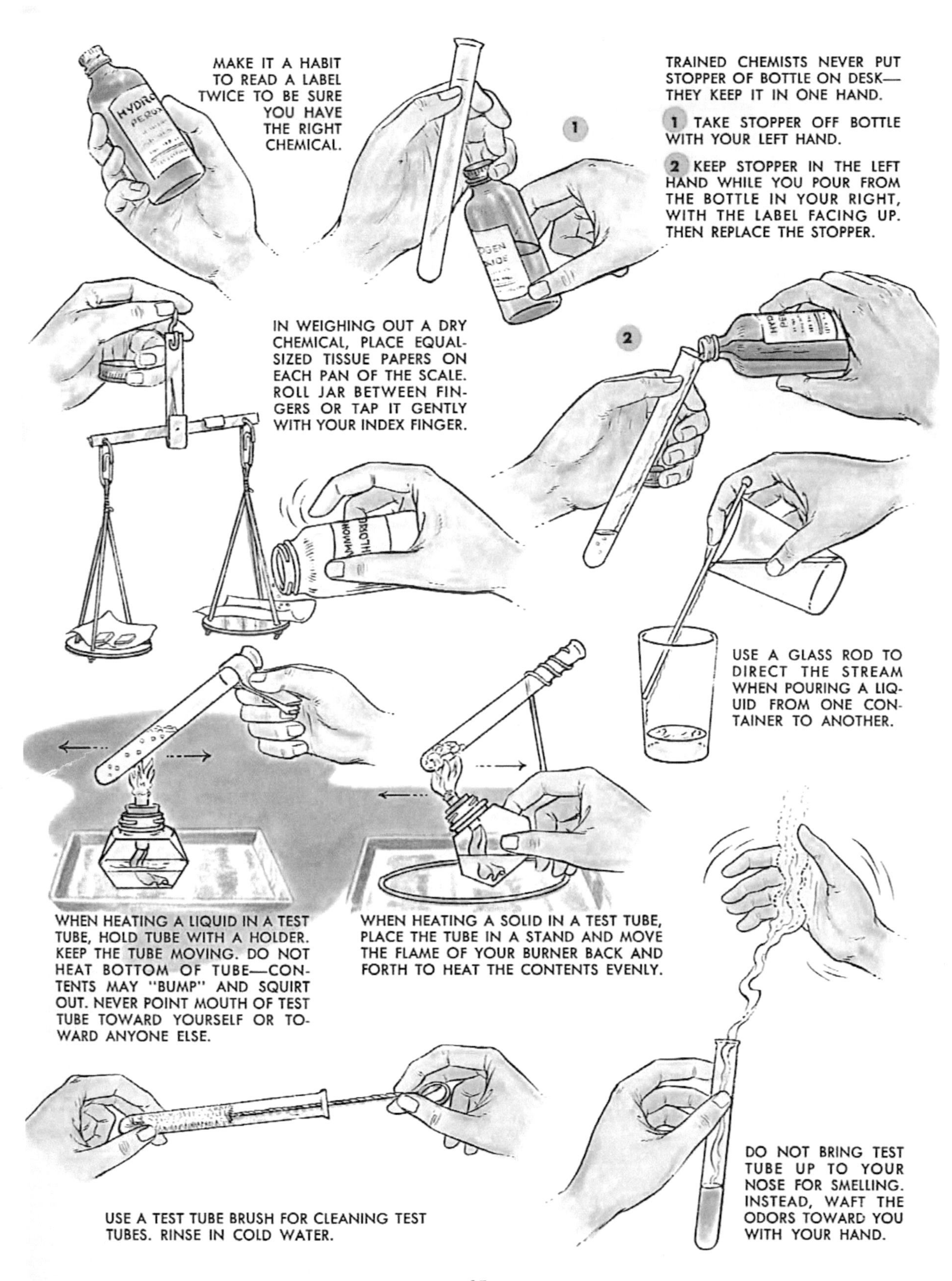
MAKE IT A HABIT TO READ A LABEL TWICE TO BE SURE YOU HAVE THE RIGHT CHEMICAL.
1
TRAINED CHEMISTS NEVER PUT STOPPER OF BOTTLE ON DESK—THEY KEEP IT IN ONE HAND.
1 TAKE STOPPER OFF BOTTLE WITH YOUR LEFT HAND.
2 KEEP STOPPER IN THE LEFT HAND WHILE YOU POUR FROM THE BOTTLE IN YOUR RIGHT, WITH THE LABEL FACING UP. THEN REPLACE THE STOPPER.
IN WEIGHING OUT A DRY CHEMICAL, PLACE EQUAL-SIZED TISSUE PAPERS ON EACH PAN OF THE SCALE. ROLL JAR BETWEEN FINGERS OR TAP IT GENTLY WITH YOUR INDEX FINGER.
2
USE A GLASS ROD TO DIRECT THE STREAM WHEN POURING A LIQUID FROM ONE CONTAINER TO ANOTHER.
WHEN HEATING A LIQUID IN A TEST TUBE, HOLD TUBE WITH A HOLDER. KEEP THE TUBE MOVING. DO NOT HEAT BOTTOM OF TUBE—CONTENTS MAY "BUMP" AND SQUIRT OUT. NEVER POINT MOUTH OF TEST TUBE TOWARD YOURSELF OR TOWARD ANYONE ELSE.
WHEN HEATING A SOLID IN A TEST TUBE, PLACE THE TUBE IN A STAND AND MOVE THE FLAME OF YOUR BURNER BACK AND FORTH TO HEAT THE CONTENTS EVENLY.
USE A TEST TUBE BRUSH FOR CLEANING TEST TUBES. RINSE IN COLD WATER.
DO NOT BRING TEST TUBE UP TO YOUR NOSE FOR SMELLING. INSTEAD, WAFT THE ODORS TOWARD YOU WITH YOUR HAND.

Mr. Faraday's Candle

In the winter of 1859, Michael Faraday, a great British scientist, gave a number of lectures for young people. The talks dealt with one subject only: the features or "phenomena" of — a candle!

"There is not a law," Faraday told his listeners, "under which any part of this universe is governed which does not come into play and is touched upon in these phenomena. There is no better, there is no more open door by which you can enter into the study of natural philosophy than by considering the phenomena of a candle." He then set out to prove his point by lighting a candle and demonstrating all the processes involved.

In burning a candle you start with a SOLID substance that turns, first, into a LIQUID, then into a GAS (or, more correctly, into a gas-like vapor). The melted candle grease is held in a level position by GRAVITY yet seems to defy gravity by rising in the wick by a force called CAPILLARY ACTION. In burning, the candle produces ENERGY in the form of LIGHT and HEAT. At the same time, it goes into CHEMICAL REACTIONS that reveal what it is made of.

As you enter the study of chemistry, you can do no better than to repeat for yourself some of the experiments that Mr. Faraday demonstrated to his young audience.

PLACE A BURNING CANDLE IN THE SUN AND CATCH THE SHADOW ON A PIECE OF WHITE PAPER. YOU WILL DISCOVER THAT IT IS THE BRIGHTEST PART OF THE FLAME THAT CASTS THE DARKEST SHADOW.

CANDLE FLAME IS BURNING VAPOR

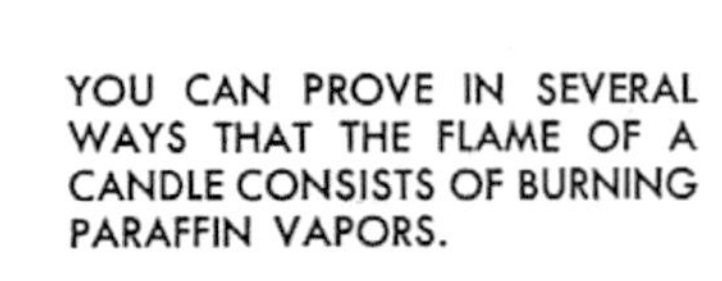

YOU CAN PROVE IN SEVERAL WAYS THAT THE FLAME OF A CANDLE CONSISTS OF BURNING PARAFFIN VAPORS.

BLOW OUT THE CANDLE, THEN QUICKLY BRING A LIGHTED MATCH INTO THE VAPORS. CANDLE IS AGAIN IGNITED.

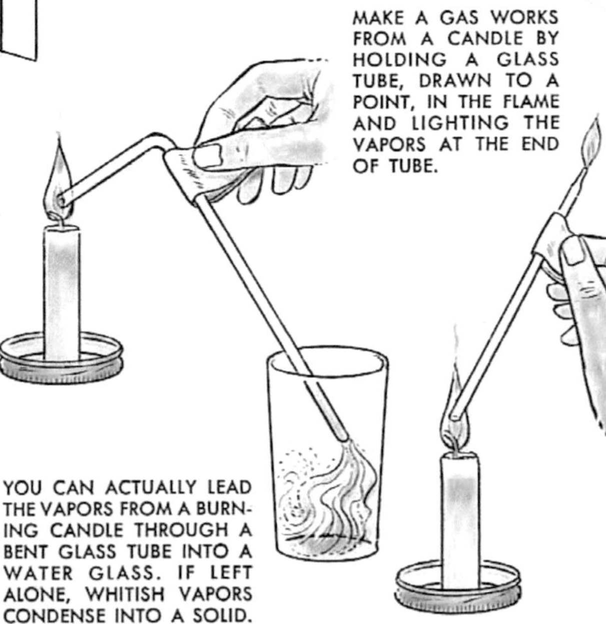

MAKE A GAS WORKS FROM A CANDLE BY HOLDING A GLASS TUBE, DRAWN TO A POINT, IN THE FLAME AND LIGHTING THE VAPORS AT THE END OF TUBE.

YOU CAN ACTUALLY LEAD THE VAPORS FROM A BURNING CANDLE THROUGH A BENT GLASS TUBE INTO A WATER GLASS. IF LEFT ALONE, WHITISH VAPORS CONDENSE INTO A SOLID.

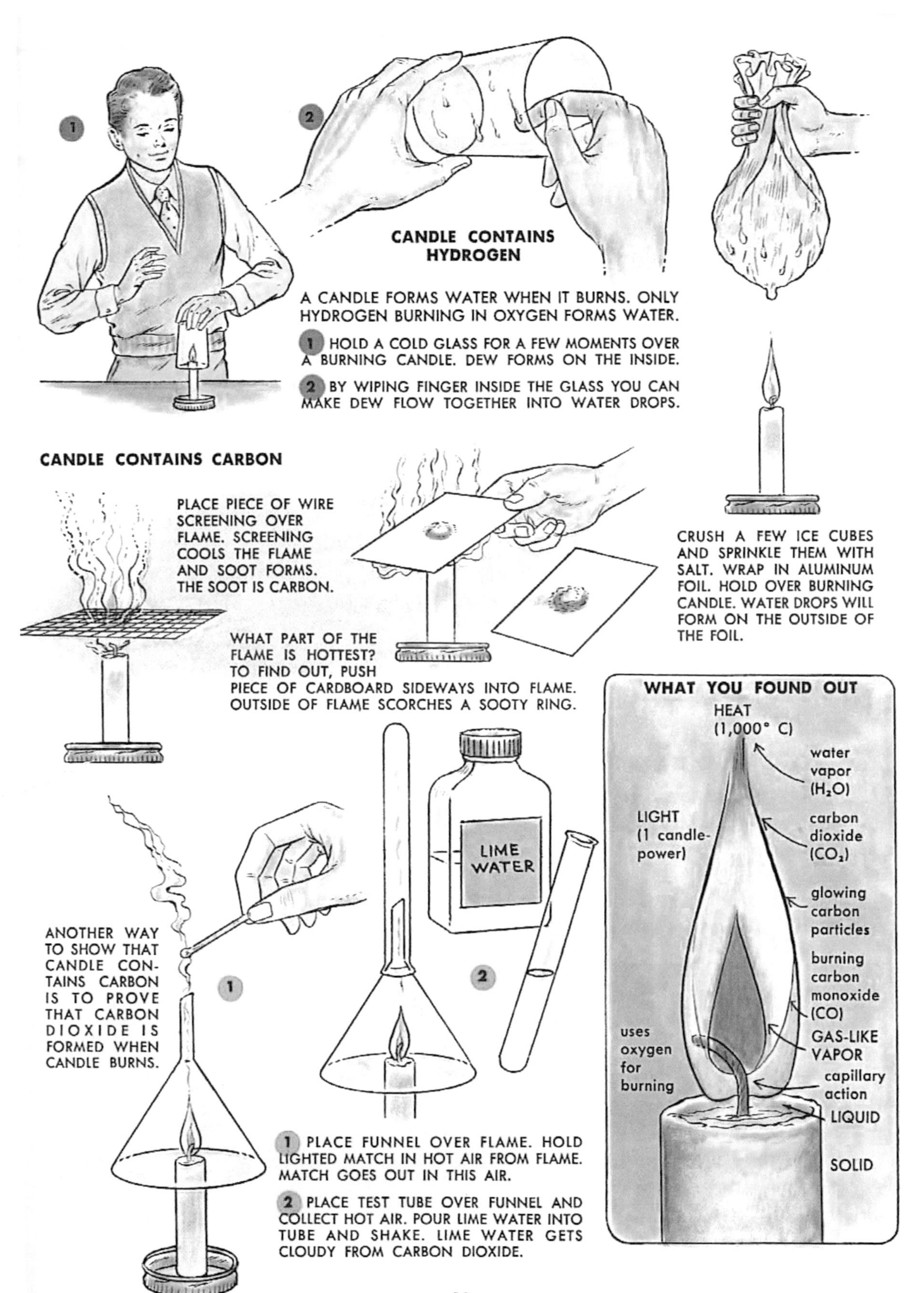

1
2
CANDLE CONTAINS HYDROGEN
A CANDLE FORMS WATER WHEN IT BURNS. ONLY HYDROGEN BURNING IN OXYGEN FORMS WATER.
1 HOLD A COLD GLASS FOR A FEW MOMENTS OVER A BURNING CANDLE. DEW FORMS ON THE INSIDE.
2 BY WIPING FINGER INSIDE THE GLASS YOU CAN MAKE DEW FLOW TOGETHER INTO WATER DROPS.
CRUSH A FEW ICE CUBES AND SPRINKLE THEM WITH SALT. WRAP IN ALUMINUM FOIL. HOLD OVER BURNING CANDLE. WATER DROPS WILL FORM ON THE OUTSIDE OF THE FOIL.
CANDLE CONTAINS CARBON
PLACE PIECE OF WIRE SCREENING OVER FLAME. SCREENING COOLS THE FLAME AND SOOT FORMS. THE SOOT IS CARBON.
WHAT PART OF THE FLAME IS HOTTEST? TO FIND OUT, PUSH PIECE OF CARDBOARD SIDEWAYS INTO FLAME. OUTSIDE OF FLAME SCORCHES A SOOTY RING.
ANOTHER WAY TO SHOW THAT CANDLE CONTAINS CARBON IS TO PROVE THAT CARBON DIOXIDE IS FORMED WHEN CANDLE BURNS.
1
LIME WATER
2
1 PLACE FUNNEL OVER FLAME. HOLD LIGHTED MATCH IN HOT AIR FROM FLAME. MATCH GOES OUT IN THIS AIR.
2 PLACE TEST TUBE OVER FUNNEL AND COLLECT HOT AIR. POUR LIME WATER INTO TUBE AND SHAKE. LIME WATER GETS CLOUDY FROM CARBON DIOXIDE.
WHAT YOU FOUND OUT
HEAT (1,000° C)
water vapor (H_2O)
LIGHT (1 candle-power)
carbon dioxide (CO_2)
glowing carbon particles
burning carbon monoxide (CO)
uses oxygen for burning
GAS-LIKE VAPOR
capillary action
LIQUID
SOLID

You—Scientist!

In 1896, a young Polish chemist, Marie Curie, and her French husband, Pierre, decided to find out why a certain uranium ore called pitchblende gave off rays that were much stronger than the uranium content of the ore could explain.

They secured a whole ton of powdered ore from a mine in northern Bohemia and set to work. First the powder had to be boiled with strong acids to extract the mysterious substance hidden in it. Then the solution had to be filtered and boiled down. What remained had to be purified by various processes which the Curies had to invent themselves.

After two years of back-breaking work they reached their goal. One night they went to the shed in which they had been working. They opened the door and stepped in without putting on the lights. All around them, the containers that held the solutions of the new substance glowed in the dark! They had discovered a new element — radium — a million times more active than uranium.

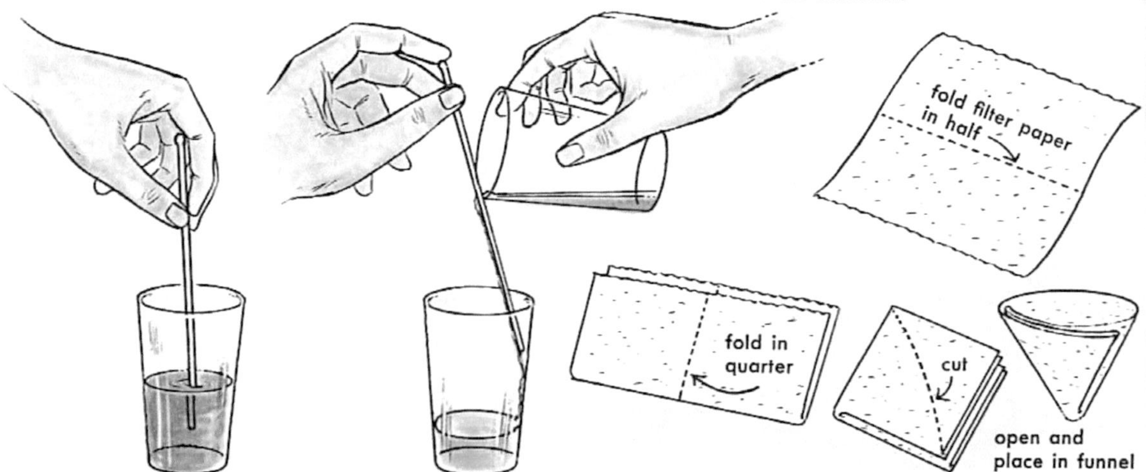

SOLUTION—STIR WATER INTO THE MIXTURE OF SALT AND DIRT. THE WATER WILL DISSOLVE THE SALT BUT NOT THE DIRT. YOU NOW HAVE THE SALT IN "WATERY SOLUTION."

DECANTATION—LET DIRT-MIXED SALT SOLUTION STAND UNTIL MOST OF THE DIRT HAS SETTLED. THEN POUR OFF THE LIQUID. THIS PROCESS IS CALLED "DECANTATION."

FILTRATION 1—THE LIQUID IS PROBABLY STILL MURKY. TO CLEAR IT, IT NEEDS TO BE FILTERED BY LETTING IT RUN THROUGH FILTER PAPER (PAPER TOWELING WILL DO).

Why tell again the story of the discovery of radium? Because it contains all the features that show the nature of the true scientist.

Curiosity first. The Curies were curious about the mystery that lay in that greyish-black powder. They became obsessed with a desire to find out — not in the hope of gaining money or fame but to establish a scientific truth.

Before starting their work, the Curies gathered all the known facts about the material with which they were to work. To this knowledge they added their own imagination, figuring out the method they had to use to arrive at the result they were seeking.

For the next two years they literally slaved in the drafty shed that was their laboratory.

After they had made their discovery, the Curies made their method of extracting radium known to the world so that other scientists could check and test what they had done.

As an example of the scientific method the Curies used, let us follow in their footsteps — but with a much simpler problem:

1 MIX THOROUGHLY ONE TABLESPOON OF DIRT AND ONE TEASPOON OF ORDINARY TABLE SALT. NOW DECIDE THAT YOU WANT TO EXTRACT THE SALT FROM THIS MIXTURE AS EARNESTLY AS THE CURIES DECIDED TO EXTRACT THE MYSTERIOUS SUBSTANCE FROM PITCHBLENDE —WITH THE EXCEPTION THAT YOU KNOW WHAT YOU ARE AFTER.

2 GET THE FACTS TOGETHER. DIRT IS "DIRTY," SALT IS WHITE. DIRT PARTICLES ARE OF MANY DIFFERENT SHAPES, SALT CONSISTS OF TINY CUBES. DIRT DOES NOT DISSOLVE IN WATER, SALT DOES.

3 NEXT FIGURE OUT A SUITABLE WAY OF SEPARATING THE TWO SUBSTANCES. ON THE BASIS OF WHAT YOU KNOW YOU SHOULD BE ABLE TO SEPARATE THEM WITH A PAIR OF TINY TWEEZERS—BUT IT WOULD PROBABLY TAKE YOU A YEAR TO DO IT. OR YOU COULD DISSOLVE THE SALT IN WATER AND SEPARATE THE SOLUTION FROM THE INSOLUBLE DIRT.

4 YOU DECIDE ON THE SECOND WAY, USING THE STEPS SHOWN ON THE BOTTOM OF THESE PAGES. IN DOING THIS, YOU DO WHAT THE CURIES DID IN EXTRACTING RADIUM AND LEARN, IN THE PROCESS, THE IMPORTANT LABORATORY TECHNIQUES OF SOLUTION, DECANTATION, FILTRATION, EVAPORATION, AND CRYSTALLIZATION.

5 FINALLY, YOU CHECK THE RESULT. THE WHITE SUBSTANCE LEFT AFTER EVAPORATION SHOULD BE SALT— BUT IS IT? IT LOOKS LIKE SALT. IT TASTES LIKE SALT. BY CHEMICAL TESTS YOU CAN PROVE THAT IT *IS* SALT.

By using the same procedure in all other experiments in this book you are learning the methods that real scientists follow in their work — you are becoming a scientist yourself.

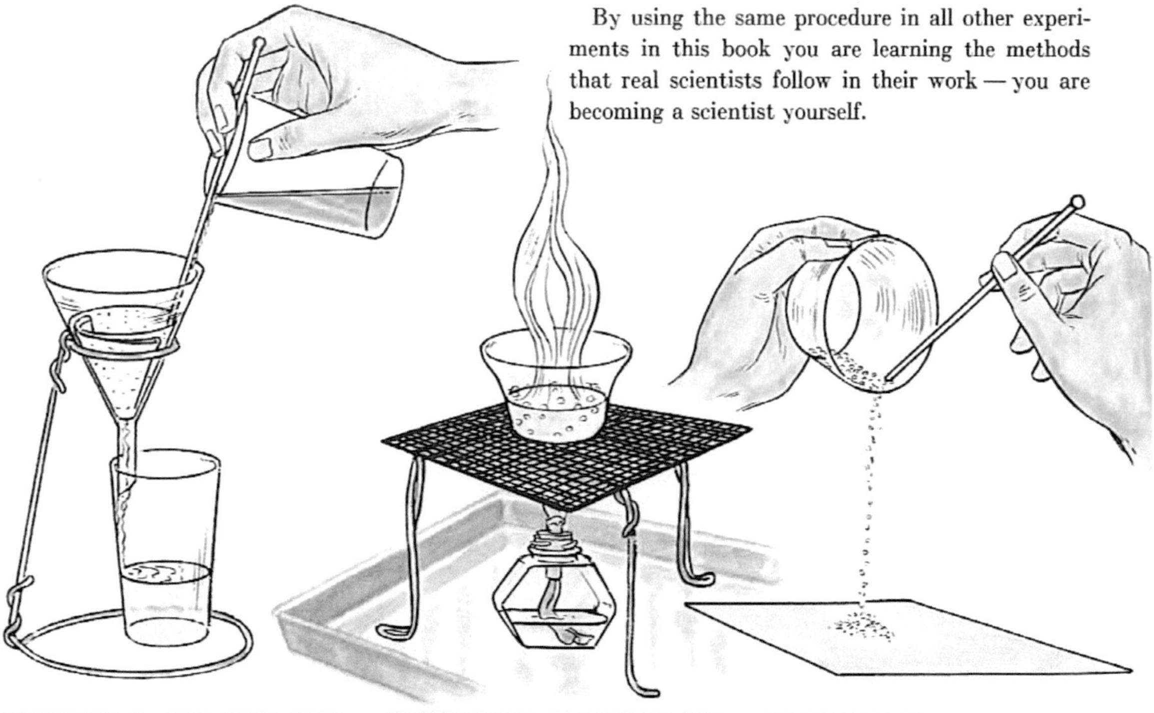

FILTRATION 2—FOLD FILTER PAPER AS SHOWN ON OPPOSITE PAGE AND FIT IT IN FUNNEL. POUR LIQUID ONTO FILTER PAPER. CLEARED LIQUID IS CALLED "FILTRATE."

EVAPORATION—THE FILTRATE CONTAINS THE SALT. THE SALT CAN NOW BE FREED BY REMOVING THE WATER BY BOILING IT AWAY. THIS IS KNOWN AS "EVAPORATION."

CRYSTALLIZATION—AS WATER IS REMOVED, THERE IS TOO LITTLE OF IT LEFT FOR THE SALT TO STAY IN SOLUTION. THE SALT MAKES ITS APPEARANCE AS TINY CRYSTALS.

Elements, Compounds, and Mixtures

In all your experiments in chemistry, you will be dealing with "matter."

Matter is anything that takes up room and has weight (or "mass"). An iron bar is matter — it takes up room and is heavy, as you very well know. Water is matter — it takes up room when you fill a pail with it, and a full pail weighs plenty. The air around you is matter — it takes up lots of room; it may not seem very heavy, yet the earth's atmosphere presses down on every square inch of your body with a weight of almost fifteen pounds.

Matter has three distinct forms. Iron, for instance, is a SOLID. Water is a LIQUID. Air has the form of a GAS.

If you should take iron and divide it again and again until you couldn't divide it any further, every tiny particle would still be iron. A thing that consists of one kind of matter only is called an ELEMENT.

Take water, on the other hand. You will learn to break water up into two kinds of matter — each of them an element. A thing in which two or more elements are combined chemically is called a COMPOUND. In a compound the proportions of the different elements that make it up are always exactly the same.

Air also consists of different kinds of matter, but they are not combined chemically — they are simply mixed together. When you make a MIXTURE, you can mix the ingredients together in any proportions that suit you.

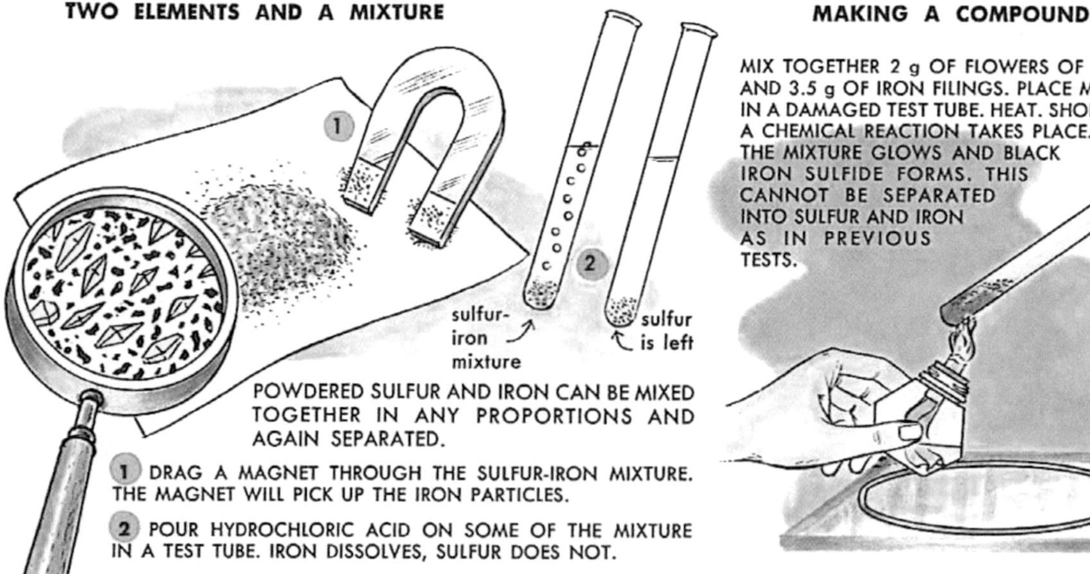

ELEMENTS ARE SUBSTANCES THAT CONSIST OF ONE KIND OF MATTER ONLY. THEY CAN BE DIVIDED INTO METALS, METALLOIDS (METAL-LIKE), NONMETALS. SEVERAL OF THE NONMETALS ARE GASES.

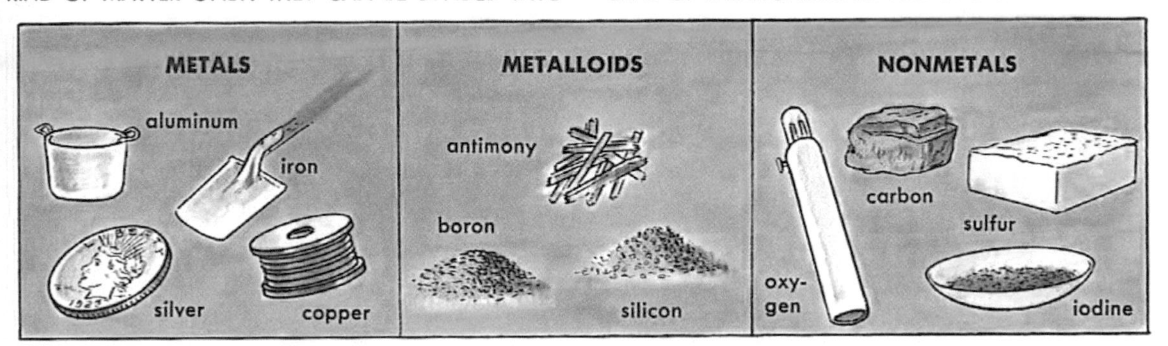

COMPOUNDS—INORGANIC. ALL COMPOUNDS CONSIST OF TWO OR MORE ELEMENTS. INORGANIC COMPOUNDS (WITH A FEW EXCEPTIONS) ARE THOSE THAT DO NOT CONTAIN THE ELEMENT CARBON.

CARBON COMPOUNDS—ORGANIC. ORIGINALLY, COMPOUNDS MADE BY LIVING THINGS (PLANTS AND ANIMALS) WERE CALLED "ORGANIC." TODAY ORGANIC CHEMISTRY COVERS THE CARBON COMPOUNDS.

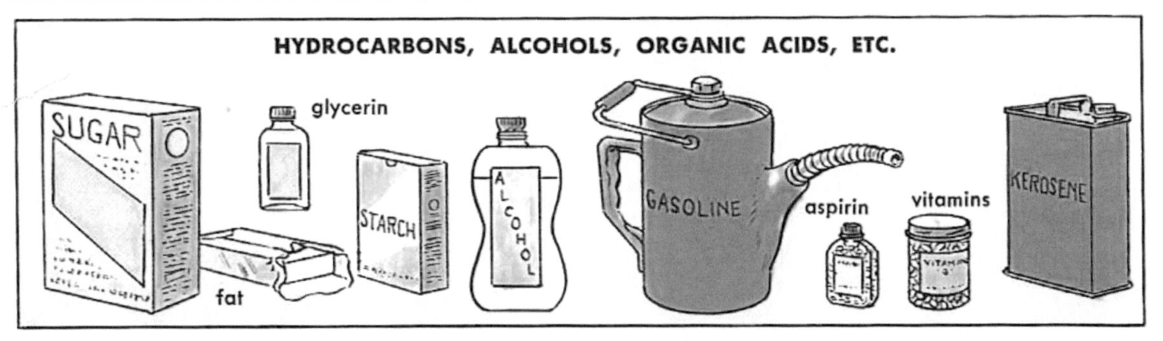

MIXTURES CAN CONSIST OF ELEMENTS OR COMPOUNDS. SOME MIXTURES ARE COARSE. SOME (COLLOIDS) CONTAIN TINY PARTICLES. STILL OTHERS (SOLUTIONS) ARE OF SAME STRUCTURE THROUGHOUT.

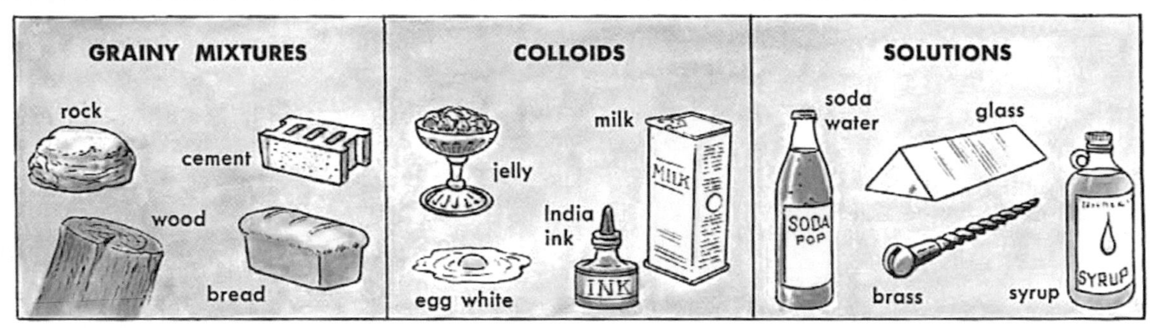

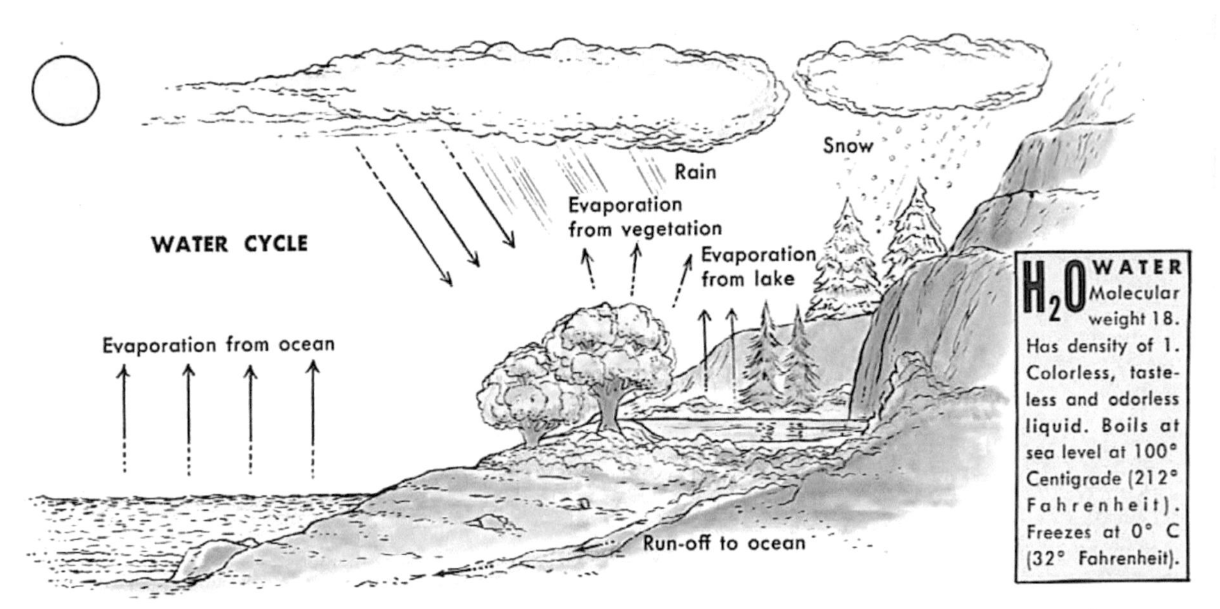

Water—Our Most Important Compound

YES, WATER is the most important of all chemical compounds. Without it, there would be no life — all human beings and all animals would thirst to death, and all plant life would wilt and die.

Fortunately, water is also the most common compound in the world. Almost three quarters of the earth's surface is covered by water. This water is forever traveling. It is turned into invisible vapor by evaporation from oceans and lakes and growing things. When cooled, the vapor forms clouds of tiny water drops. Further cooling makes the drops fall to earth as rain or snow that fill up rivers and lakes and oceans and continue the water cycle.

Chemists use nature's method to produce chemically pure water. They turn ordinary tap water into steam by boiling, then turn the steam back into water by cooling. This process is called distillation and the water is called distilled water.

WATER AS A SOLVENT

THE MOST IMPORTANT FUNCTION OF WATER IN CHEMICAL EXPERIMENTS IS AS A SOLVENT—THAT IS, A LIQUID IN WHICH CHEMICALS MAY BE DISSOLVED. FIND OUT BY AN EASY EXPERIMENT WHETHER HEATING THE WATER HELPS IN DISSOLVING A CHEMICAL.

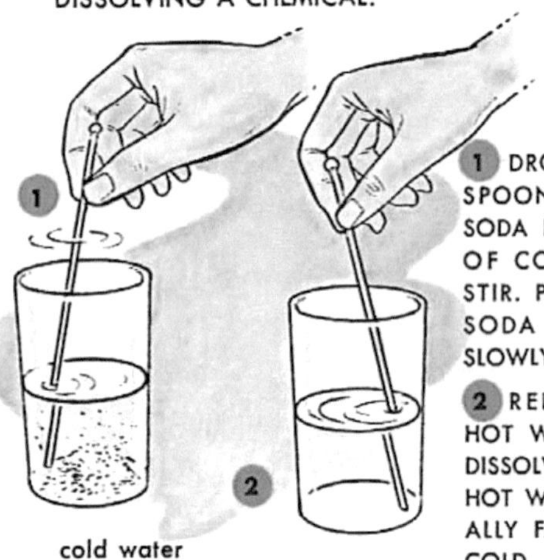

1 DROP 1 TABLESPOON WASHING SODA IN ½ GLASS OF COLD WATER. STIR. PART OF THE SODA DISSOLVES SLOWLY.

2 REPEAT WITH HOT WATER. SODA DISSOLVES QUICKLY. HOT WATER IS USUALLY FASTER THAN COLD FOR PREPARING A SOLUTION.

WATER AS A CATALYST

WATER HELPS BRING ABOUT MANY CHEMICAL REACTIONS WITHOUT ITSELF ENTERING INTO THEM. A SUBSTANCE THAT ACTS THIS WAY IS CALLED A CATALYST.

1 PLACE 1 TEASPOON DRY BAKING POWDER IN SMALL JAR. ATTACH WIRE TO CANDLE. LIGHT CANDLE AND LOWER IT INTO JAR. CANDLE GOES ON BURNING.

2 NOW POUR WARM WATER ON THE BAKING POWDER. A CHEMICAL REACTION MAKES THE POWDER FOAM. THE GAS RELEASED IS CARBON DIOXIDE. IT MAKES CANDLE FLAME FLICKER AND GO OUT.

ELECTROLYSIS OF WATER

ELECTRICITY CAN BE USED TO BREAK WATER APART INTO THE TWO ELEMENTS OF WHICH IT CONSISTS —THE GASES HYDROGEN AND OXYGEN.

YOU CAN GET THE REQUIRED ELECTRICITY FROM THREE OR FOUR ORDINARY FLASHLIGHT BATTERIES. YOU WILL ALSO NEED TWO PIECES OF INSULATED COPPER WIRE AND TWO "ELECTRODES" MADE FROM CARBON RODS.

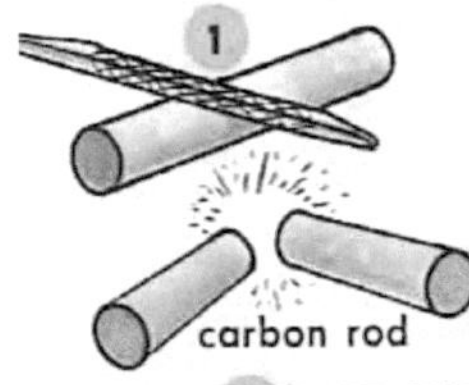

Making Electrodes

1 SCORE THE MIDDLE OF THE CARBON ROD FROM AN OLD FLASHLIGHT BATTERY, USING A FILE. BREAK THE ROD INTO TWO PIECES.

2 BARE THE WIRE FOR 2" AT EACH END OF TWO 18" LENGTHS OF INSULATED WIRE. TIE ONE BARED WIRE AROUND END OF EACH OF CARBON ROD HALVES.

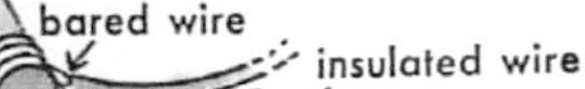

3 BIND ELECTRICIAN'S TAPE FIRMLY AROUND CARBON RODS SO THAT NO WIRE IS EXPOSED.

electrician's tape

paper clip

Setting up Electrolysis

WATER IS A POOR CONDUCTOR OF ELECTRICITY—SO YOU DISSOLVE 1 TABLESPOON OF WASHING SODA IN 1 PINT OF WATER AND FILL A WATER GLASS AND TWO TEST TUBES WITH THIS SOLUTION. THEN SET UP THE APPARATUS AS SHOWN AT RIGHT.

MATERIALS FOR EXPERIMENTS

AN ORDINARY FLASHLIGHT BATTERY WILL GIVE YOU MATERIALS YOU NEED FOR EXPERIMENTS ON THIS AND SEVERAL FOLLOWING PAGES.

1 OPEN UP BATTERY CASE CAREFULLY WITH A CAN OPENER AND CLEAN THE ZINC CASING.

2 SCRAPE CARBON ROD CLEAN WITH DULL KNIFE.

3 DRY OUT THE MOIST BLACK POWDER, WHICH IS MOSTLY MANGANESE DIOXIDE. STORE IN JAR. THROW REMAINING PARTS OF THE BATTERY AWAY.

Performing the Electrolysis

1 SLIP THE TOP OF A CARBON ELECTRODE UP INTO EACH OF THE TWO TEST TUBES.

2 BIND THREE—OR, BETTER, FOUR—FLASHLIGHT BATTERIES TOGETHER WITH ADHESIVE TAPE, TOP OF ONE TOUCHING BOTTOM OF THE NEXT.

3 WITH ADHESIVE TAPE FASTEN THE BARED END OF THE WIRE LEADING FROM ONE CARBON ROD ELECTRODE TO THE TOP OF THE FIRST BATTERY.

4 TAPE THE BARED END OF THE WIRE FROM THE OTHER ELECTRODE TO BOTTOM OF LAST BATTERY.

AS SOON AS CONNECTION IS MADE, AIR BUBBLES BEGIN TO COLLECT IN THE TWO TEST TUBES—ABOUT TWICE AS FAST IN ONE AS IN THE OTHER.

adhesive tape

flashlight batteries

TEST FOR HYDROGEN

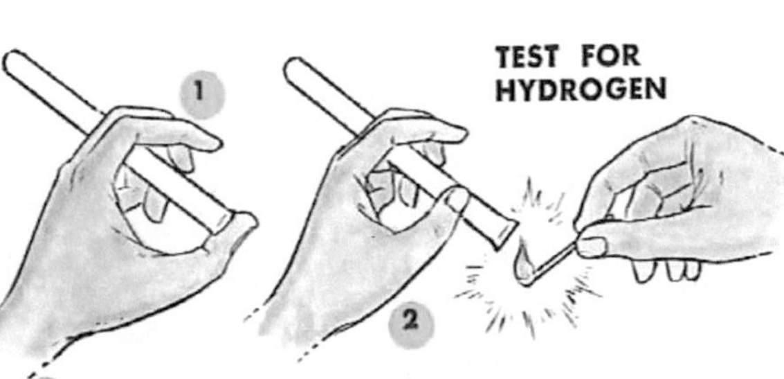

1 WITH YOUR THUMB, CLOSE THE MOUTH OF THE TEST TUBE FIRST FILLED WITH GAS. LIFT THE TUBE OUT OF THE WATER, MOUTH DOWN.

2 BRING LIGHTED MATCH TO THE MOUTH OF THE TUBE. CONTENTS BURN WITH A SOFT "POP!" THIS IS THE TEST FOR HYDROGEN.

TEST FOR OXYGEN

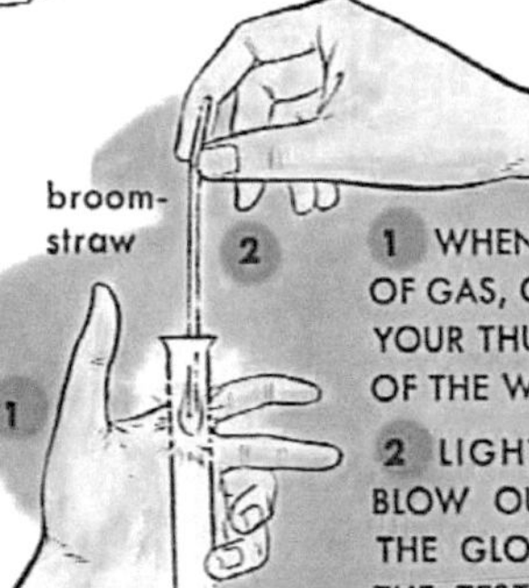

1 WHEN SECOND TUBE IS FULL OF GAS, CLOSE ITS MOUTH WITH YOUR THUMB. LIFT THE TUBE OUT OF THE WATER WITH MOUTH UP.

2 LIGHT A BROOMSTRAW. BLOW OUT THE FLAME. BRING THE GLOWING END DOWN IN THE TEST TUBE. GLOWING EMBER BURSTS INTO BRIGHT FLAME. THIS IS TEST FOR OXYGEN.

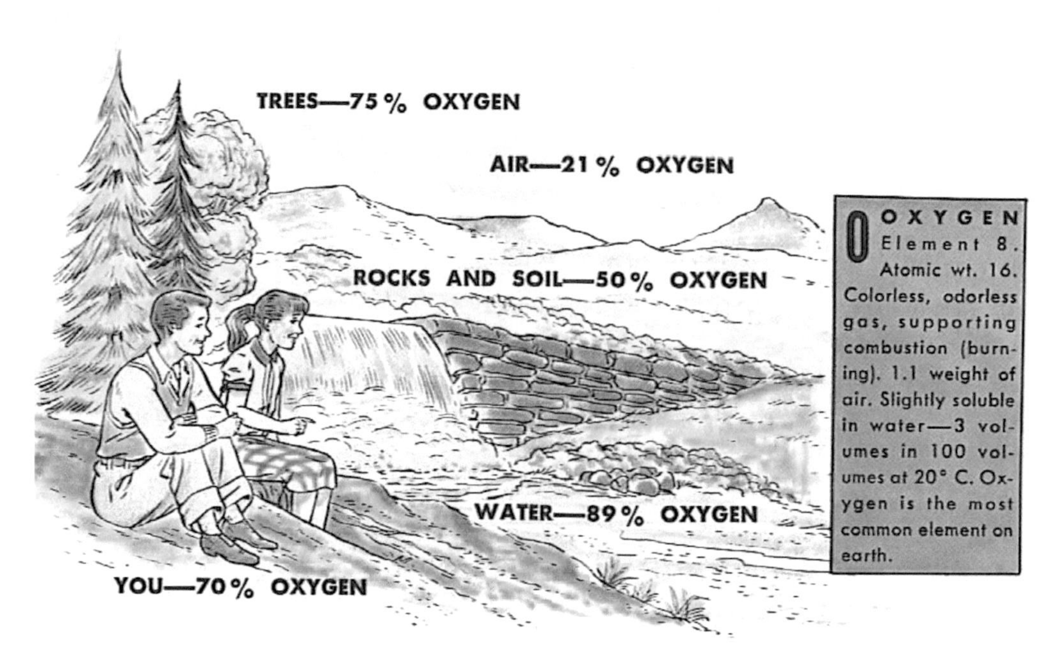

O OXYGEN Element 8. Atomic wt. 16. Colorless, odorless gas, supporting combustion (burning). 1.1 weight of air. Slightly soluble in water—3 volumes in 100 volumes at 20° C. Oxygen is the most common element on earth.

Oxygen—The Breath of Life

IF YOU could hold your breath for a few minutes so that no air could get into your lungs, you would die.

For thousands of years, people have known that no human being can live without air. But it was not until Karl Scheele, a Swedish chemist, in 1772, and Joseph Priestley, an Englishman, in 1774, discovered and described oxygen that people knew that it is the oxygen in the air that is important to life.

Both of these scientists discovered that things burn more fiercely in pure oxygen than they do in the mixture of oxygen and other gases called "air."

In the lab, oxygen is produced by driving it out of certain oxygen-containing compounds. A good one to use in the home lab is hydrogen peroxide. You can get it at a drug store in a 3% solution. Hydrogen peroxide is related to water.

Water, as you know, consists of 2 parts of hydrogen to 1 part of oxygen. You could write it: Hydrogen 2 — Oxygen 1. That's pretty much what chemists do — except that they abbreviate the names to initials, use small numbers, and don't bother about the number 1. The formula becomes H_2O.

Hydrogen peroxide contains 2 parts of hydrogen to every 2 parts of oxygen. How would you write it in chemical language?

H_2O_2? You're perfectly right!

H_2O_2 becomes water (H_2O) and gives off oxygen (O) when you throw a catalyst into it. For a catalyst, you can use the manganese dioxide from an old flashlight battery (page 25).

IT'S A LONG STEP FROM THE DISCOVERY OF OXYGEN IN 1772 TO ITS PRESENT-DAY USE IN INDUSTRY AND HOSPITALS, AIRPLANES AND SPACE SHIPS, AND FOR SENDING SATELLITES INTO ORBIT.

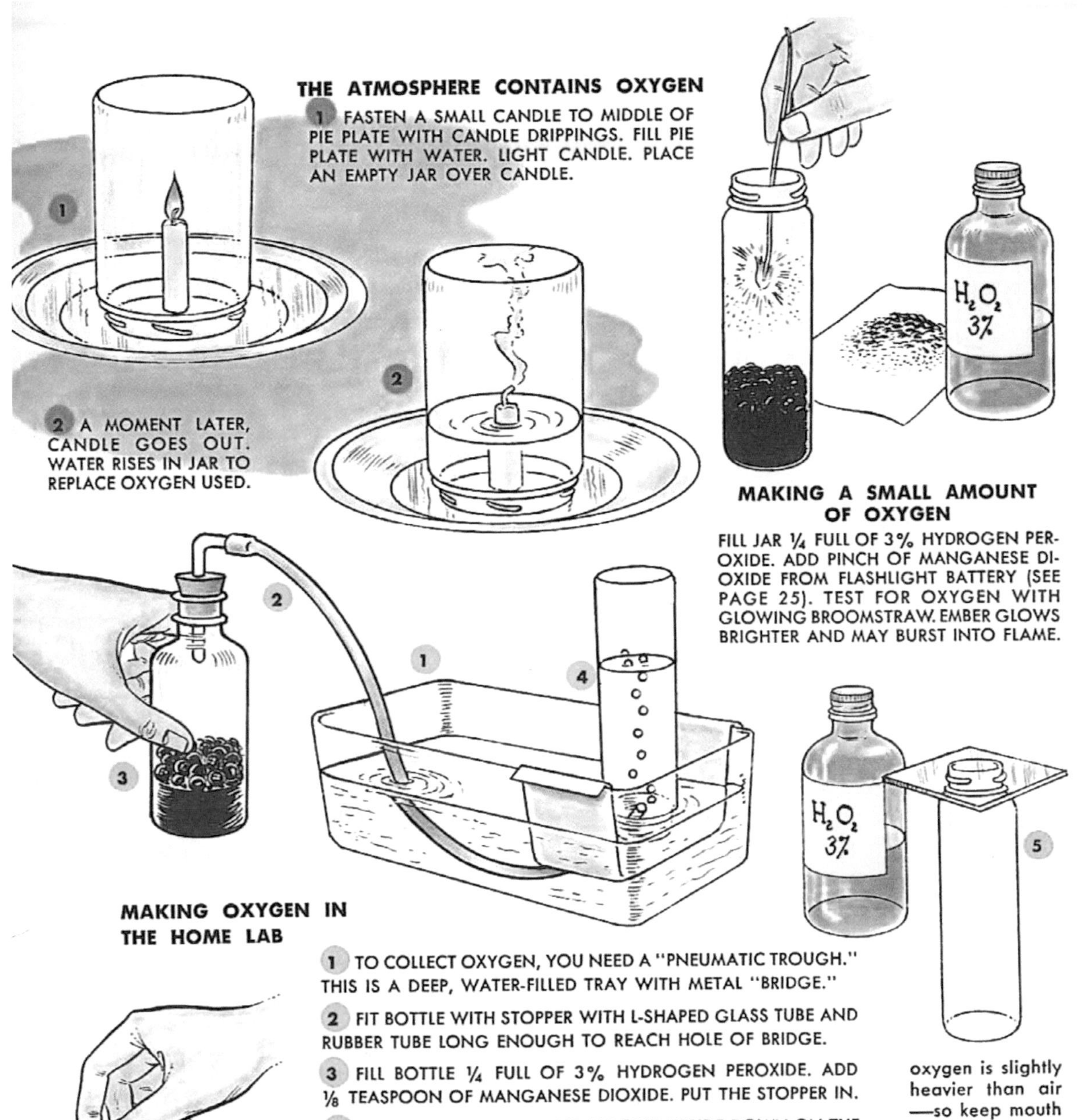

THE ATMOSPHERE CONTAINS OXYGEN

1 FASTEN A SMALL CANDLE TO MIDDLE OF PIE PLATE WITH CANDLE DRIPPINGS. FILL PIE PLATE WITH WATER. LIGHT CANDLE. PLACE AN EMPTY JAR OVER CANDLE.

2 A MOMENT LATER, CANDLE GOES OUT. WATER RISES IN JAR TO REPLACE OXYGEN USED.

MAKING A SMALL AMOUNT OF OXYGEN

FILL JAR ¼ FULL OF 3% HYDROGEN PEROXIDE. ADD PINCH OF MANGANESE DIOXIDE FROM FLASHLIGHT BATTERY (SEE PAGE 25). TEST FOR OXYGEN WITH GLOWING BROOMSTRAW. EMBER GLOWS BRIGHTER AND MAY BURST INTO FLAME.

MAKING OXYGEN IN THE HOME LAB

1 TO COLLECT OXYGEN, YOU NEED A "PNEUMATIC TROUGH." THIS IS A DEEP, WATER-FILLED TRAY WITH METAL "BRIDGE."

2 FIT BOTTLE WITH STOPPER WITH L-SHAPED GLASS TUBE AND RUBBER TUBE LONG ENOUGH TO REACH HOLE OF BRIDGE.

3 FILL BOTTLE ¼ FULL OF 3% HYDROGEN PEROXIDE. ADD ⅛ TEASPOON OF MANGANESE DIOXIDE. PUT THE STOPPER IN.

4 FILL JAR WITH WATER AND PLACE IT UPSIDE DOWN ON THE BRIDGE IN SUCH A WAY THAT THE OXYGEN BUBBLES INTO IT AND FILLS IT BY FORCING OUT AND REPLACING THE WATER

5 WHEN JAR IS FULL OF OXYGEN, SLIDE A GLASS PLATE UNDER OPENING (OR PUT STOPPER IN IT). TURN JAR RIGHT SIDE UP—QUICKLY, TO PREVENT THE OXYGEN FROM ESCAPING.

oxygen is slightly heavier than air —so keep mouth of jar up

MANY MATERIALS BURN IN OXYGEN

1 ATTACH TUFT OF STEEL WOOL TO WIRE. HEAT TO RED HEAT OVER ALCOHOL BURNER. LOWER INTO JAR OF OXYGEN. IRON BURSTS INTO FLAME.

2 PLACE SMALL PIECE OF SULFUR IN CROOK OF BENT STRIP OF TIN CUT FROM CAN. IGNITE SULFUR WITH MATCH. LOWER INTO JAR OF OXYGEN. SULFUR BURNS WITH A BRILLIANT, BLUE LIGHT.

"BRIDGE" FOR "PNEUMATIC TROUGH" MADE FROM 2½" STRIP OF TIN CAN.

Hydrogen—Lightest of All

HYDROGEN is the lightest element in existence—1/14 the weight of air. For this reason one of its early uses was for filling balloons. The first man-carrying gas balloon was sent up by the Frenchman, Jacques Charles, in 1783. The danger of using an explosive gas for this purpose was demonstrated in 1937 in the *Hindenburg* disaster, when the hydrogen-filled Zeppelin dirigible exploded on arriving at Lakehurst, New Jersey, after a trip across the Atlantic Ocean. Thirty-six people lost their lives.

Hydrogen is one of the most important of all the elements. It is found in all living things — your own body is approximately 10 per cent hydrogen. Water, as you know, is part hydrogen. So is the food you eat, the milk you drink, the clothes you wear, and such common, everyday things as gasoline and fuel oil and cooking gas.

In the home lab, you can make hydrogen by adding strips of zinc from a flashlight battery to hydrochloric acid which consists of hydrogen (H) and another gas called chlorine (Cl). The zinc forms a compound ($ZnCl_2$) with the chlorine and sets the hydrogen free (H_2).

HENRY CAVENDISH, WHO DISCOVERED HYDROGEN IN 1766, HAD NO IDEA OF THE ASTONISHING FORCE OF HYDROGEN WHEN RELEASED IN A BOMB.

HYDROGEN FORMS WATER WHEN IT BURNS

1 FILL TEST TUBE ¾ FULL OF HYDRO-CHLORIC ACID. ADD A COUPLE OF ZINC STRIPS. BUBBLES OF HYDROGEN FORM IMMEDIATELY.

2 CLOSE TEST TUBE WITH RUBBER STOPPER WITH GLASS TUBE DRAWN TO JET POINT. COVER APPARATUS WITH A TOWEL.

3 PLACE EMPTY TEST TUBE OVER GLASS TUBE. AFTER 1 MINUTE, TEST THIS TUBE FOR HYDROGEN WITH LIGHTED MATCH. IF TUBE "BARKS," PUT IT BACK. AFTER ANOTHER MINUTE, TRY AGAIN. WHEN SOFT "POP" TELLS YOU GAS IS PURE, LIGHT JET.

4 HOLD A COLD GLASS OVER HYDROGEN FLAME. DEW COVERING THE INSIDE OF THE GLASS SHOWS THAT WATER IS FORMING.

HYDROGEN IS LIGHTEST GAS KNOWN

1 FILL A POP BOTTLE ¼ FULL OF HALF-AND-HALF MIXTURE OF HYDROCHLORIC ACID AND WATER. DROP IN HALF A DOZEN ZINC STRIPS. LET NO FLAME COME NEAR!

2 FIT BALLOON ON MOUTH OF BOTTLE.

3 WHEN BALLOON IS INFLATED, TIE OPENING WITH STRING AND REMOVE FROM BOTTLE. IF PERMITTED, BALLOON WILL RISE TO CEILING INDOORS. OUTDOORS, IT WILL SOAR UP IN THE SKY.

MAKING HYDROGEN IN THE LAB

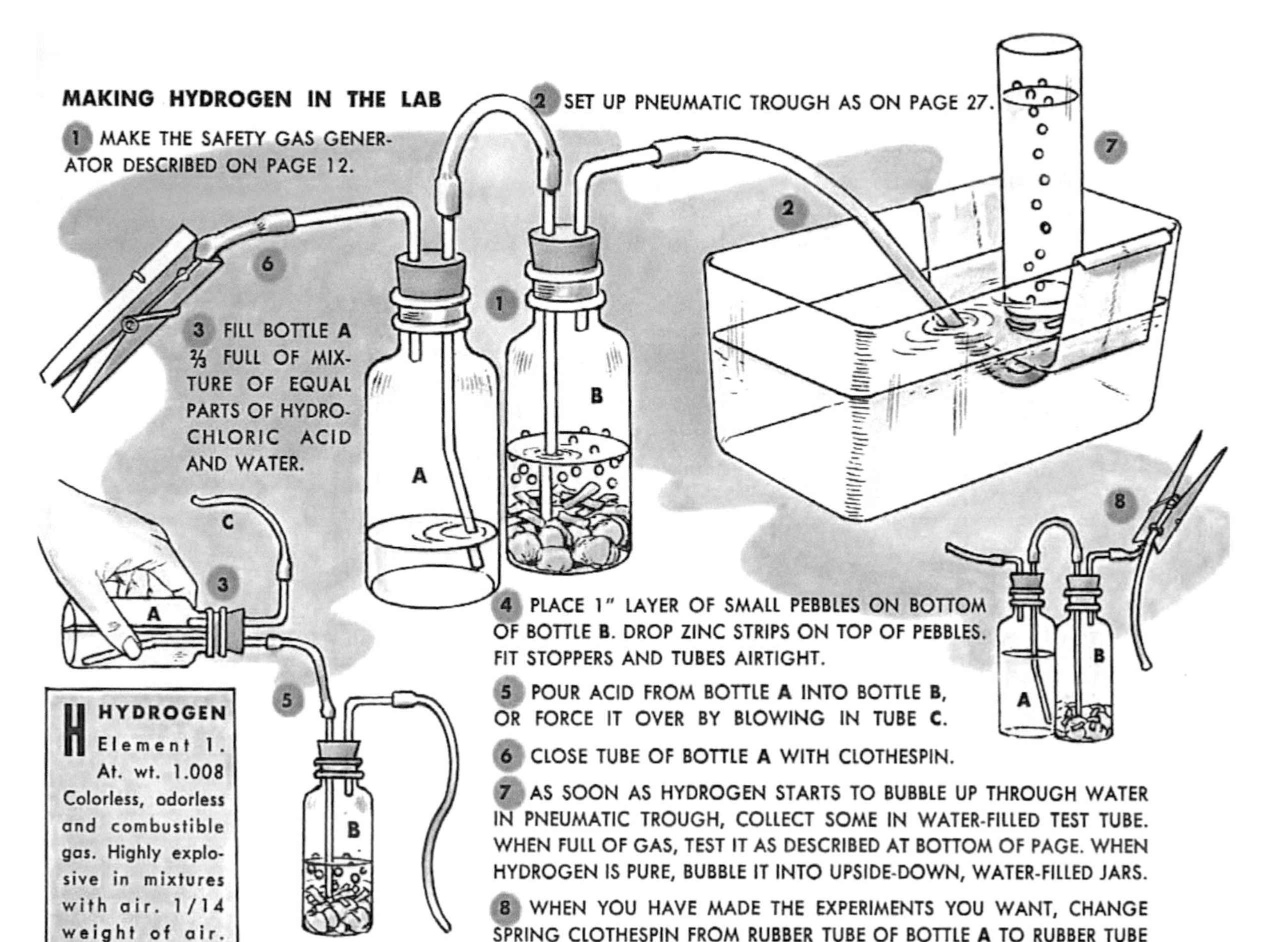

1 MAKE THE SAFETY GAS GENERATOR DESCRIBED ON PAGE 12.

2 SET UP PNEUMATIC TROUGH AS ON PAGE 27.

3 FILL BOTTLE **A** ⅔ FULL OF MIXTURE OF EQUAL PARTS OF HYDROCHLORIC ACID AND WATER.

4 PLACE 1" LAYER OF SMALL PEBBLES ON BOTTOM OF BOTTLE **B**. DROP ZINC STRIPS ON TOP OF PEBBLES. FIT STOPPERS AND TUBES AIRTIGHT.

5 POUR ACID FROM BOTTLE **A** INTO BOTTLE **B**, OR FORCE IT OVER BY BLOWING IN TUBE **C**.

6 CLOSE TUBE OF BOTTLE **A** WITH CLOTHESPIN.

7 AS SOON AS HYDROGEN STARTS TO BUBBLE UP THROUGH WATER IN PNEUMATIC TROUGH, COLLECT SOME IN WATER-FILLED TEST TUBE. WHEN FULL OF GAS, TEST IT AS DESCRIBED AT BOTTOM OF PAGE. WHEN HYDROGEN IS PURE, BUBBLE IT INTO UPSIDE-DOWN, WATER-FILLED JARS.

8 WHEN YOU HAVE MADE THE EXPERIMENTS YOU WANT, CHANGE SPRING CLOTHESPIN FROM RUBBER TUBE OF BOTTLE **A** TO RUBBER TUBE OF BOTTLE **B**. HYDROGEN FORCES ACID FROM BOTTLE **B** BACK INTO **A**. WHEN ACID NO LONGER TOUCHES ZINC, REACTION STOPS.

H **HYDROGEN**
Element 1.
At. wt. 1.008
Colorless, odorless and combustible gas. Highly explosive in mixtures with air. 1/14 weight of air. Slightly soluble in water—1.8 volumes in 100 volumes at 20° C.

PLAYING SAFE WITH HYDROGEN

IN MIXTURES WITH AIR, HYDROGEN IS HIGHLY EXPLOSIVE. FOLLOW SAFETY RULES BELOW.

- MAKE ONLY SMALL AMOUNTS OF HYDROGEN IN THE HOME LAB. A 4-OZ. GENERATOR BOTTLE WILL GIVE YOU ALL THE HYDROGEN YOU NEED. MAKE ALL CONNECTIONS AIRTIGHT.
- TEST HYDROGEN FOR PURITY BY COLLECTING A TEST TUBE FULL OF IT AND BRINGING A LIGHTED MATCH TO MOUTH OF TUBE, AS SHOWN ON PAGE 25. HYDROGEN MIXED WITH AIR EXPLODES WITH A SHARP "BARK." PURE HYDROGEN BURNS WITH A QUIET "POP."
- KEEP FLAME AWAY FROM YOUR MAIN GENERATOR BOTTLE.
- IGNITE HYDROGEN ONLY FROM TEST TUBE GENERATOR DESCRIBED ON OPPOSITE PAGE, AND THEN ONLY AFTER YOU HAVE TESTED IT FOR PURITY.

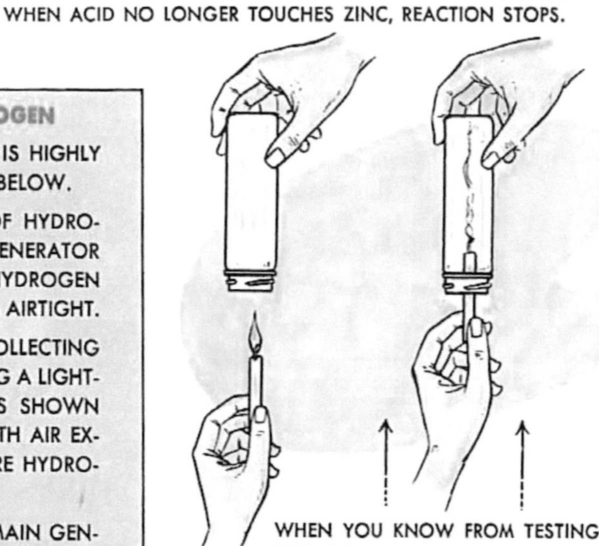

WHEN YOU KNOW FROM TESTING SAMPLES OF GAS COLLECTED IN TEST TUBES THAT HYDROGEN IS PURE, FILL SMALL JAR WITH IT. LIFT JAR OUT OF WATER, MOUTH DOWN. BRING LIGHTED CANDLE UP INTO JAR. HYDROGEN BURNS AT MOUTH OF JAR. CANDLE GOES OUT.

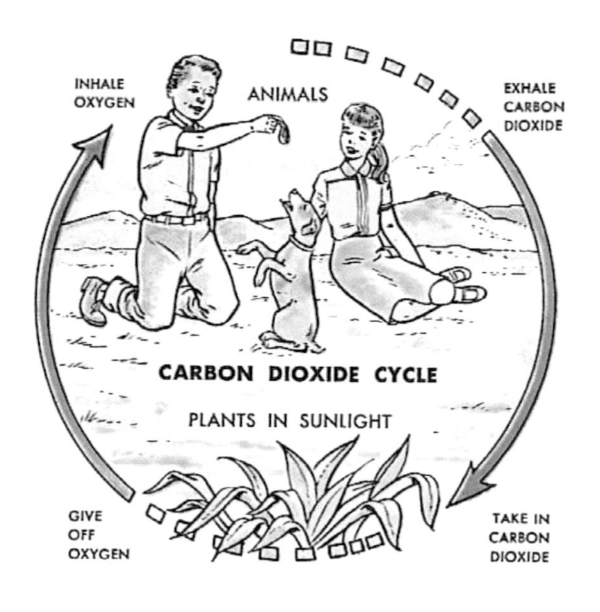

Carbon Dioxide

You have already learned in experimenting with a burning candle that when something containing carbon burns in the air, a gas, carbon dioxide (CO_2), is formed. This is one of the most important gases for human life. The reason is that green plants, in sunlight, are able to take the carbon out of the carbon dioxide in the air and, by combining it with oxygen and hydrogen from water and with various minerals in the soil, produce all the vegetable matter that humans and animals eat.

You cannot see the CO_2 in the air — but you can see it when it has been cooled and compressed into a solid block of "dry ice." When dissolved in water (H_2O), carbon dioxide (CO_2) forms a weak acid (H_2CO_3). You know the taste of this acid from soda water — the bubbles are CO_2 being set free.

Carbonic acid combines with many metals to make "carbonates." You can drive the CO_2 out of most carbonates with the help of a weak acid — even with vinegar, which is diluted acetic acid.

LIQUID CARBON DIOXIDE IS USED IN FIRE EXTINGUISHERS.

MAKING A FIRE EXTINGUISHER MODEL

1 PUSH A SHORT GLASS TUBE WITH A JET TIP INTO A RUBBER STOPPER. WRAP BICARBONATE OF SODA IN A SHEET OF TOILET TISSUE. ATTACH SODA PACKAGE TO TUBE WITH A RUBBER BAND.

2 FILL BOTTLE HALF FULL OF MIXTURE OF 1 PART VINEGAR AND 1 PART WATER. PUT IN THE STOPPER.

3 HOLD STOPPER FIRMLY IN PLACE WITH TWO FINGERS. TURN BOTTLE UPSIDE DOWN. THE CO_2 FORMED BY MIXING VINEGAR AND SODA DRIVES WATER OUT IN POWERFUL JET.

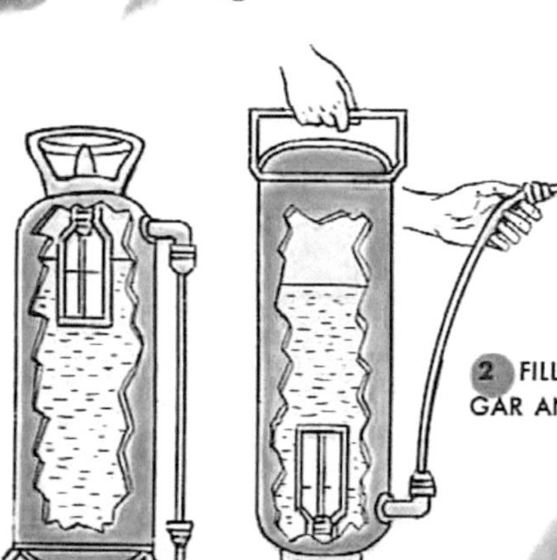

CHEMICAL FIRE EXTINGUISHERS CONTAIN SOLUTION OF BAKING SODA AND A BOTTLE OF SULFURIC ACID. WHEN TURNED UPSIDE DOWN, THE CHEMICALS MIX AND FORM CARBON DIOXIDE WHICH FORCES OUT THE WATER.

CO_2 CARBON DIOXIDE Compound.
Molecular wt. 44. Colorless, odorless gas. Does not burn. Does not support combustion (burning). 1.529 weight of air. Fairly soluble in water—88 volumes in 100 volumes at 20° C.

FEATURES OF CARBON DIOXIDE

CO_2 IS HEAVIER THAN AIR AND DOES NOT SUPPORT BURNING. YOU CAN PROVE BOTH POINTS:

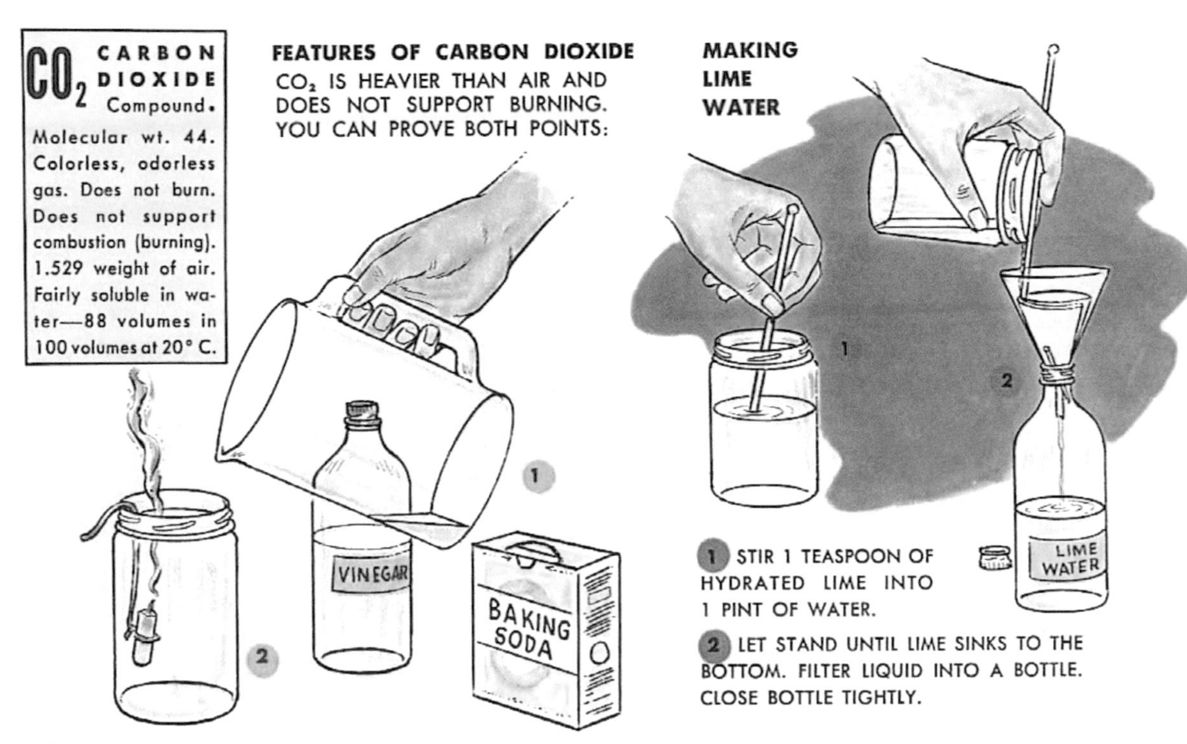

MAKING LIME WATER

1 STIR 1 TEASPOON OF HYDRATED LIME INTO 1 PINT OF WATER.

2 LET STAND UNTIL LIME SINKS TO THE BOTTOM. FILTER LIQUID INTO A BOTTLE. CLOSE BOTTLE TIGHTLY.

1 PLACE 1 TEASPOON OF BAKING SODA IN A PITCHER. POUR A SMALL AMOUNT OF WHITE VINEGAR OVER THE SODA.

2 HANG A LIGHTED CANDLE IN A JAR BY A WIRE. POUR THE CARBON DIOXIDE FORMED IN THE PITCHER INTO THE JAR THE WAY YOU WOULD POUR WATER. WHEN THE CARBON DIOXIDE REACHES THE TOP OF THE CANDLE, THE FLAME GOES OUT.

BURNING PRODUCES CARBON DIOXIDE

HANG BURNING CANDLE IN JAR CONTAINING A FEW ml LIME WATER. COVER TOP WITH A GLASS PLATE. WHEN CANDLE HAS GONE OUT, SHAKE LIME WATER UP WITH THE AIR. MILKINESS PROVES THAT CO_2 HAS BEEN PRODUCED.

MAKING CO_2 FROM MARBLE

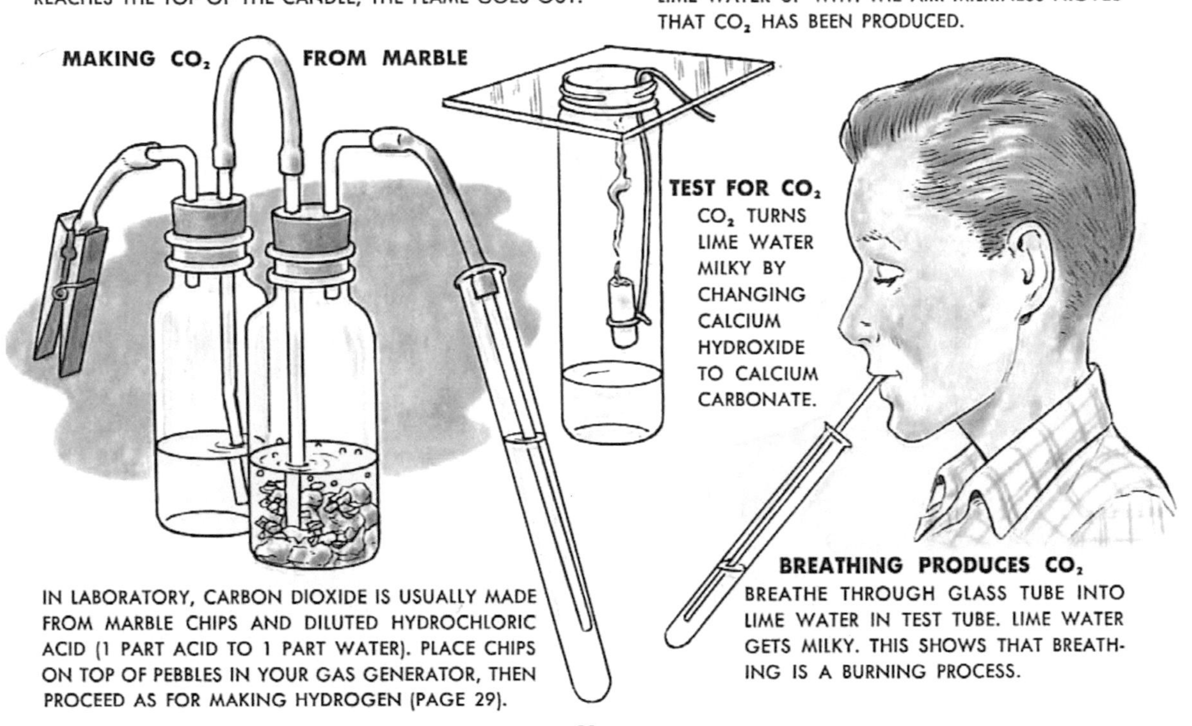

IN LABORATORY, CARBON DIOXIDE IS USUALLY MADE FROM MARBLE CHIPS AND DILUTED HYDROCHLORIC ACID (1 PART ACID TO 1 PART WATER). PLACE CHIPS ON TOP OF PEBBLES IN YOUR GAS GENERATOR, THEN PROCEED AS FOR MAKING HYDROGEN (PAGE 29).

TEST FOR CO_2

CO_2 TURNS LIME WATER MILKY BY CHANGING CALCIUM HYDROXIDE TO CALCIUM CARBONATE.

BREATHING PRODUCES CO_2

BREATHE THROUGH GLASS TUBE INTO LIME WATER IN TEST TUBE. LIME WATER GETS MILKY. THIS SHOWS THAT BREATHING IS A BURNING PROCESS.

NITROGEN GOES INTO BOTH FERTILIZERS AND EXPLOSIVES.

DRY AMMONIA GAS IS USED IN THE LARGE-SCALE PRODUCTION OF ICE.

N NITROGEN Element 7. At. wt. 14.008. Colorless, odorless gas. Does not burn. Does not support combustion (burning). .967 weight of air. Slightly soluble in water—1.5 volumes in 100 vols. at 20°C.

Nitrogen and Its Compounds

When you burn anything in the air, only about one-fifth of the air goes into chemical combination with what you are burning. The rest (except for a small fraction) does not enter into the process. It is a gas called nitrogen (N) — the most abundant free element on earth.

Nitrogen is what you might call a "lazy" element. It does not help in burning nor does it burn if you try to ignite it. It is only at high temperatures and under great pressures that a chemist can make nitrogen combine with another element, hydrogen, to form ammonia gas (NH_3), from which other nitrogen compounds can be made.

Yet, in nature, tiny bacteria on the roots of certain plants can take nitrogen from the air and make it combine with oxygen and minerals in the soil into "nitrates." And that is of tremendous importance to all of us — for all plants need nitrates if they are to thrive. If plants do not get nitrates naturally, the farmer must add them to his soil in the form of some kind of fertilizer.

You will not have much satisfaction out of working with nitrogen itself, but you will find it interesting to deal with some of its compounds — especially with ammonia gas (NH_3). You will also want to have a look at one of the half dozen combinations nitrogen makes with oxygen, the brown gas called nitrogen dioxide (NO_2).

NITROGEN FROM THE ATMOSPHERE

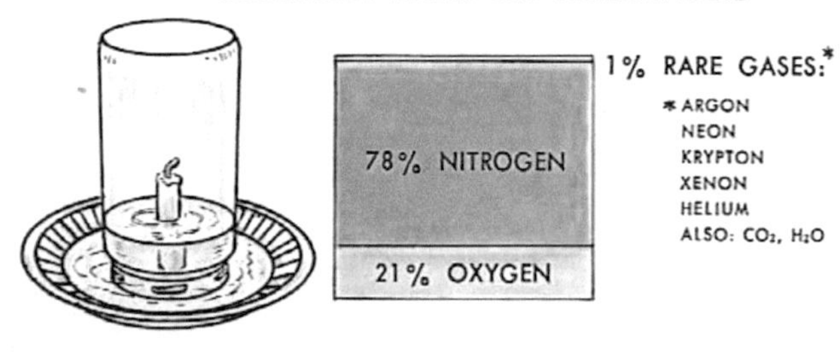

REPEAT CANDLE-BURNING EXPERIMENT ON PAGE 27. UNUSED GAS IS ALMOST ALL NITROGEN—WITH SMALL PERCENTAGE OF RARE GASES AND CARBON DIOXIDE.

NITROGEN DIOXIDE

IN A WELL-VENTILATED ROOM, HEAT EQUAL AMOUNTS OF SALTPETER AND SODIUM BISULFATE IN DRY TEST TUBE. IN A MOMENT, A BROWN GAS FORMS. IT IS NITROGEN DIOXIDE. DO NOT INHALE —GAS IS VERY IRRITATING.

NH_3 AMMONIA Compound. Molecular weight 17. Colorless gas with strong, penetrating odor. .596 weight of air. Highly soluble in water—70,000 vols. in 100 vols. at 20°C.

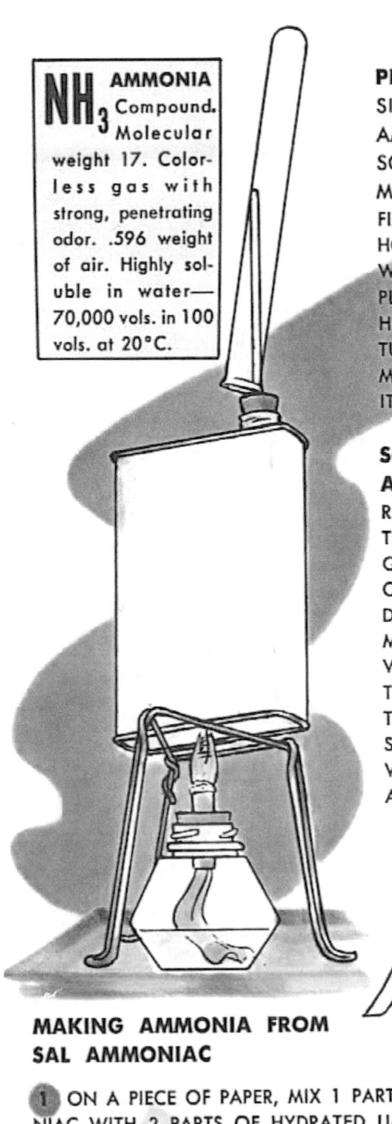

PRODUCING AMMONIA

SIMPLEST WAY OF PRODUCING AMMONIA IS TO GET IT FROM ITS SOLUTION AS HOUSEHOLD AMMONIA.

FILL PINT CAN ONE QUARTER FULL OF HOUSEHOLD AMMONIA. FIT STOPPER WITH 6" GLASS TUBE IN OPENING. PLACE TEST TUBE OVER GLASS TUBE. HEAT CAN OVER LOW FLAME. TEST TUBE IS FULL OF AMMONIA WHEN MOIST, RED LITMUS PAPER HELD AT ITS MOUTH TURNS BLUE.

SOLUBILITY OF AMMONIA

REMOVE A FILLED TEST TUBE FROM GAS GENERATOR CAN, MOUTH DOWN. CLOSE MOUTH OF TUBE WITH THUMB. OPEN TUBE UNDER WATER. AMMONIA DISSOLVES EASILY, WATER RUSHES IN AND FILLS TUBE.

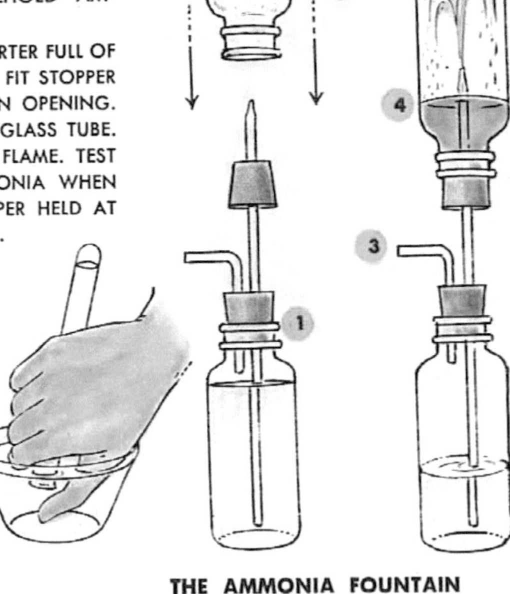

THE AMMONIA FOUNTAIN

AMMONIA'S EXTRAORDINARY SOLUBILITY CAN BE SHOWN IN A SPECTACULAR DEMONSTRATION.

1 MAKE UP APPARATUS AS SHOWN IN ILLUSTRATION. FILL IT WITH WATER. ADD 5 DROPS OF PHENOLPHTHALEIN SOLUTION.

2 FILL DRY, EMPTY BOTTLE WITH AMMONIA FROM GENERATOR CAN. KEEPING BOTTLE UPSIDE DOWN, PLACE IT FIRMLY ON TOP STOPPER OF APPARATUS.

3 BLOW INTO L-SHAPED GLASS TUBE TO DRIVE A FEW DROPS OF WATER UP INTO THE UPPER BOTTLE.

4 SUDDENLY, WATER SPURTS FROM LOWER BOTTLE UP INTO UPPER BOTTLE IN A FOUNTAIN THAT TURNS PINK AS AMMONIA REACTS ON PHENOLPHTHALEIN.

MOIST, RED LITMUS PAPER TURNS BLUE IN AMMONIA.

MAKING AMMONIA FROM SAL AMMONIAC

1 ON A PIECE OF PAPER, MIX 1 PART OF SAL AMMONIAC WITH 2 PARTS OF HYDRATED LIME. ADD A FEW DROPS OF WATER. DROP MIXTURE INTO A TEST TUBE. PROVIDE TUBE WITH STOPPER AND L-SHAPED GLASS TUBE. THEN HEAT OVER LOW FLAME.

2 COLLECT AMMONIA IN DRY TEST TUBE. TEST IT WITH LITMUS PAPER AND FOR SOLUBILITY.

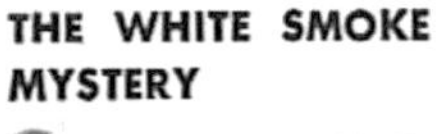

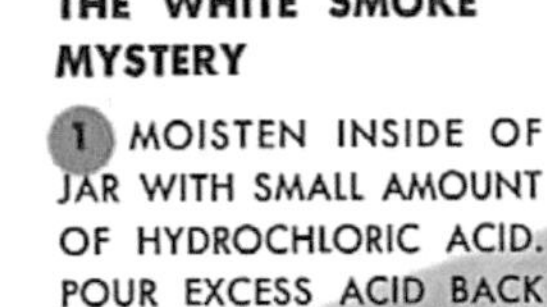

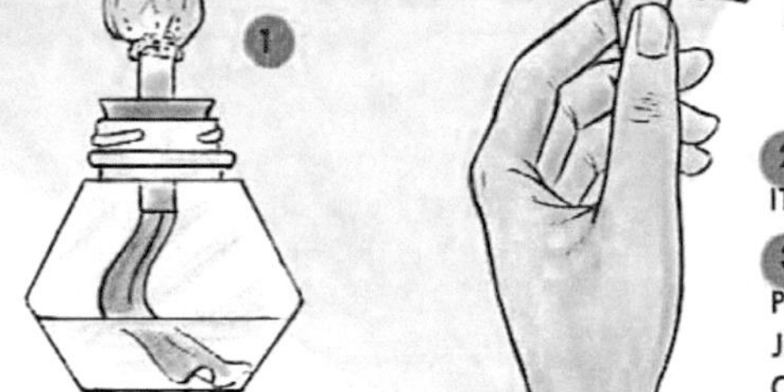

THE WHITE SMOKE MYSTERY

1 MOISTEN INSIDE OF JAR WITH SMALL AMOUNT OF HYDROCHLORIC ACID. POUR EXCESS ACID BACK INTO ITS BOTTLE. COVER JAR WITH SQUARE OF CARDBOARD.

2 FILL ANOTHER JAR WITH AMMONIA. PLACE IT UPSIDE DOWN ON CARDBOARD.

3 HOLD ON TO AMMONIA-FILLED JAR AND PULL CARDBOARD AWAY. IMMEDIATELY, BOTH JARS FILL WITH "SMOKE" OF TINY AMMONIUM CHLORIDE CRYSTALS.

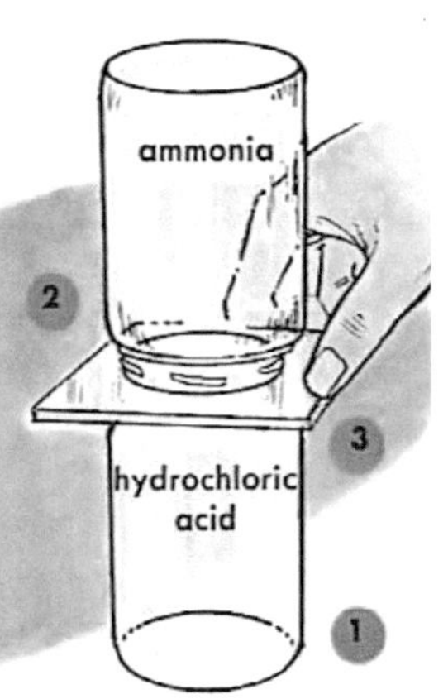

Chlorine—Friend and Foe

CHLORINE IS a gas of great importance. We wouldn't be certain of safe drinking water in our cities if it weren't for chlorine — a small amount of it in the water kills the dangerous germs that may lurk in it. Chlorine is also used extensively in bleaching.

Chlorine is a friendly gas when it is used correctly. But it is dangerous when used improperly because it affects the lungs. As a "poison gas" it caused many casualties in World War I.

You can produce chlorine as a greenish-yellow gas by driving it out of one of its compounds — hydrochloric acid (HCl), which consists of hydrogen (H) and chlorine (Cl), or a common laundry bleach ("Clorox" or others), which is a solution of sodium hypochlorite ($NaClO$).

Have a bottle of diluted household ammonia (90% water, 10% household ammonia) on hand. Sniff this if you get too strong a whiff of chlorine.

NOTE: Perform these experiments out-of-doors or before an open window. Be careful not to breathe fumes.

MAKING TEST PAPER FOR CHLORINE

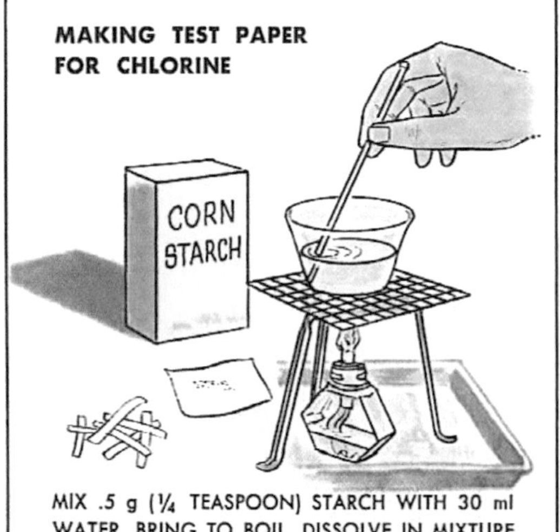

MIX .5 g (1/4 TEASPOON) STARCH WITH 30 ml WATER. BRING TO BOIL. DISSOLVE IN MIXTURE A SMALL AMOUNT OF POTASSIUM IODIDE (AS MUCH AS TWO GRAINS OF RICE). DIP STRIPS OF FILTER PAPER IN MIXTURE; THEN DRY THEM.

CHLORINE FROM HYDROCHLORIC ACID

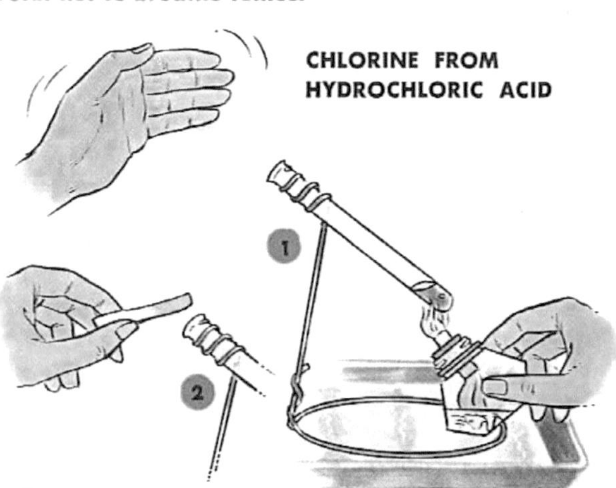

1 Put .5 g (1/8 TEASPOON) MANGANESE DIOXIDE INTO TEST TUBE. ADD 3 ml (1/8 TEST TUBE) UNDILUTED HYDROCHLORIC ACID. HEAT GENTLY. CHLORINE FORMS. WAFT A LITTLE CAREFULLY TOWARD YOU FOR A SNIFF.

2 TEST GAS BY HOLDING MOISTENED STARCH-IODIDE PAPER AT MOUTH OF TUBE. PAPER TURNS BLUE.

MAKING CHLORINE IN THE HOME LAB

1 MAKE APPARATUS SHOWN AT RIGHT. POUR 1 INCH OF LIQUID BLEACH (CLOROX) INTO BOTTLE **A**. BOTTLE **B** IS EMPTY. BOTTLE **C** HAS WATER IN WHICH ½ TEASPOON LYE IS DISSOLVED.

2 TAKE STOPPER OUT OF BOTTLE **A**. DROP IN ½ TEASPOON SODIUM BISULFATE (SANI-FLUSH). REPLACE THE STOPPER.

3 CHLORINE GAS FORMS AND FILLS **B**.

4 LYE WATER IN BOTTLE **C** ABSORBS EXCESS OF CHLORINE GAS.

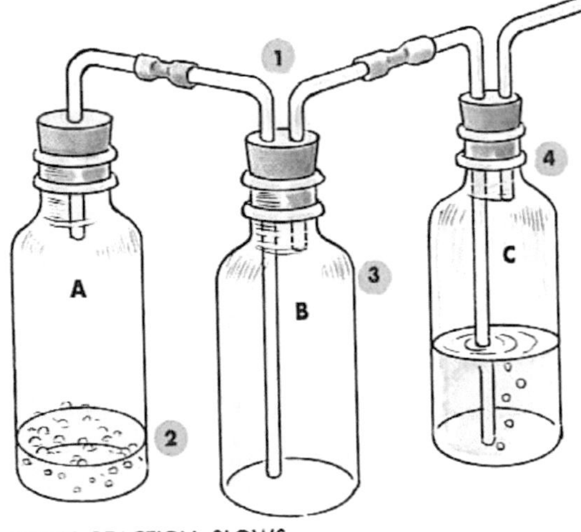

WHEN REACTION SLOWS ADD MORE SODIUM BISULFATE

EXPERIMENTS WITH CHLORINE

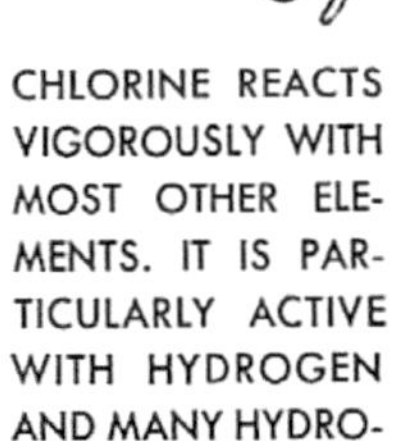

CHLORINE REACTS VIGOROUSLY WITH MOST OTHER ELEMENTS. IT IS PARTICULARLY ACTIVE WITH HYDROGEN AND MANY HYDROGEN COMPOUNDS.

LOWER A BURNING CANDLE INTO A BOTTLE OF CHLORINE GAS. A DENSE SMOKE OF CARBON IS FORMED. THE CHLORINE COMBINES WITH THE HYDROGEN OF THE CANDLE AND SETS THE CARBON IN IT FREE AS SOOT.

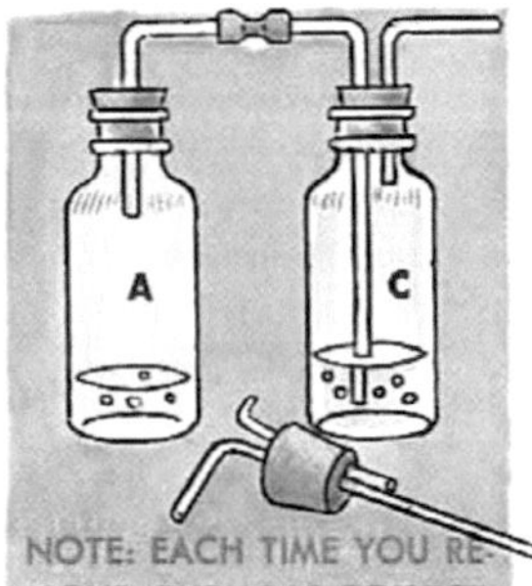

NOTE: EACH TIME YOU REMOVE THE GAS-COLLECTING BOTTLE **B** FOR EXPERIMENT, CONNECT BOTTLES **A** AND **C** TO PREVENT CHLORINE FROM GETTING OUT IN THE ROOM.

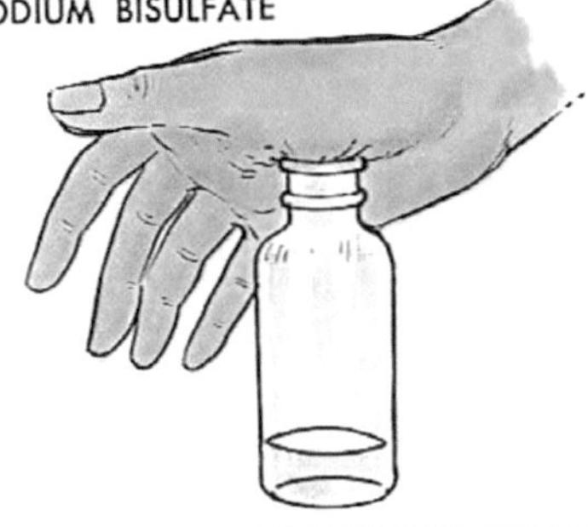

TO SHOW THE SOLUBILITY OF CHLORINE, POUR A SMALL AMOUNT OF WATER INTO A CHLORINE-FILLED BOTTLE. CLOSE THE BOTTLE MOUTH WITH YOUR PALM. SHAKE. THE CHLORINE DISSOLVES AND THE BOTTLE STICKS TO YOUR PALM FROM THE SUCTION CREATED.

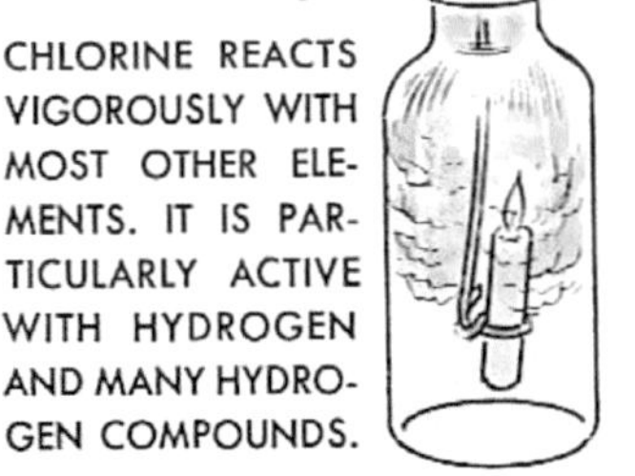

CHLORINE WILL COMBINE DIRECTLY WITH SEVERAL METALS. IRON ACTUALLY BURNS IN CHLORINE GAS!

FASTEN A SMALL WAD OF STEEL WOOL TO A PIECE OF WIRE. HEAT IT WITH A MATCH AND LOWER IT INTO CHLORINE-FILLED BOTTLE. A HEAVY BROWN SMOKE OF IRON CHLORIDE POURS OUT.

CHLORINE HAS GREAT USE IN BLEACHING COTTON AND LINEN AND WOOD PULP. YET IT IS NOT THE CHLORINE THAT PERFORMS THE BLEACHING.

1 FILL A BOTTLE WITH CHLORINE GAS. HANG IN IT (FROM A CORK OR FROM A PIECE OF CARDBOARD) A STRIP OF DRY, BRIGHTLY COLORED COTTON CLOTH. NOTHING HAPPENS. COLOR OF CLOTH IS NOT AFFECTED.

2 MOISTEN THE CLOTH AND AGAIN HANG IT IN THE CHLORINE. SOON THE COLORS FADE—ONLY TRULY "FAST" COLORS REMAIN. CHLORINE, IN CONTACT WITH WATER, COMBINES WITH THE HYDROGEN AND LIBERATES OXYGEN. THE LIBERATED OXYGEN DOES THE BLEACHING.

THE ALCHEMISTS USED FANCIFUL FIGURES TO REPRESENT THE CHEMICALS WITH WHICH THEY WORKED.

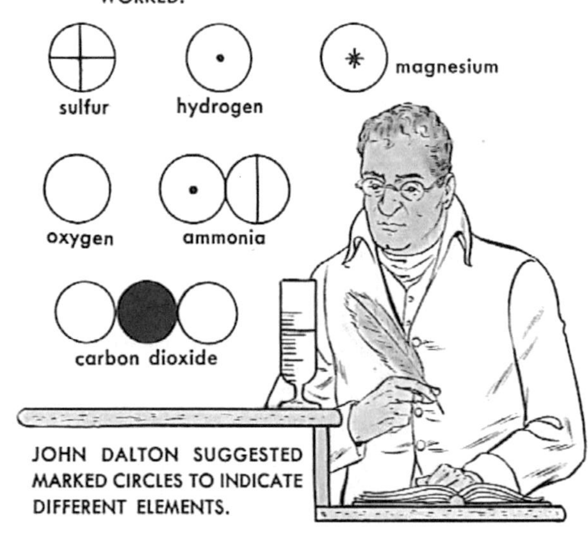

JOHN DALTON SUGGESTED MARKED CIRCLES TO INDICATE DIFFERENT ELEMENTS.

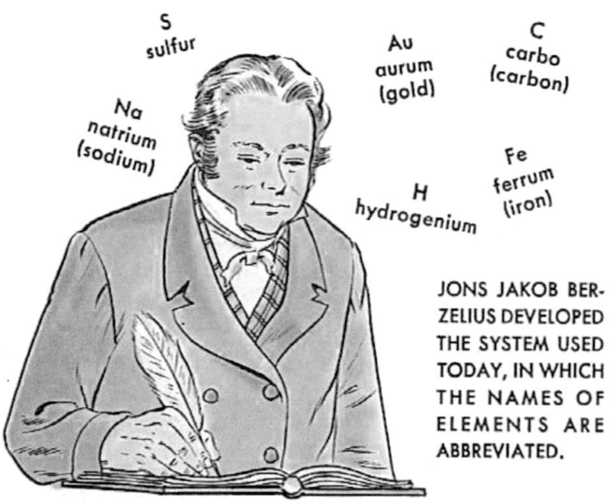

JONS JAKOB BERZELIUS DEVELOPED THE SYSTEM USED TODAY, IN WHICH THE NAMES OF ELEMENTS ARE ABBREVIATED.

Chemical Shorthand

So FAR YOU have experimented with oxygen and hydrogen, carbon dioxide and nitrogen, and chlorine; you have also separated water into the two elements of which it consists, and have combined the two elements iron and sulfur into a chemical compound. In taking notes of your experiments you are certain to have learned that it is much quicker to write "H" than "hydrogen," and easier to write "CO_2" than "carbon dioxide." Before long, it will seem the simplest and most logical thing in the world to use these abbreviations of the names of the different elements rather than the full names.

Yet it took chemists hundreds of years before they settled on this uniform method of writing out their chemical formulas.

In the early days of chemistry no one bothered to do much writing about it. But it became necessary for the alchemists to write down their experiments — how else could they retrace their steps in case they actually hit upon the gold they were seeking? They invented a whole line of complicated symbols that only they could understand.

As chemists delved deeper and deeper into the mysteries of matter it became more and more important for them to write out their experiments in such a way that all other chemists would know what they were trying to explain.

The first to invent a usable system was John Dalton, an English scientist. The invention was almost forced upon him.

In his study of chemistry he had become convinced that all chemical reactions could be explained in terms of the tiniest possible part of one element reacting with the tiniest possible part of another. These particles he called "atoms." The smallest possible part of the compound that resulted he called a "compound atom"— today we call it a "molecule."

To explain his "atomic theory" Dalton made use of circles, each with a marking to indicate a specific element. These circles served to explain Dalton's theory but they were too difficult to work with to show complicated chemical reactions.

A Swedish chemist, Jons Jakob Berzelius, worked out a simpler system — the same system scientists use today.

For his symbols he took the first letter of the Latin name of each element — C for "carbo," S for "sulfur." Where two names started with the same letter, he added a small letter to one of the symbols to distinguish the two elements from each other — he used Ca for "calcium," for instance, to distinguish it from carbon (C).

But Berzelius went an important step further.

By then the French chemist, Joseph Louis Proust, had discovered that whenever elements form compounds these are always of a very definite composition — the "Law of Definite Composition." Water molecules, for example, always contain the same number of hydrogen and oxygen atoms. And Dalton had found that when two elements combine in different ways they do this in simple proportions — the "Law of Multiple Proportions." One atom of carbon and one atom of oxygen make carbon monoxide; one atom of carbon and two atoms of oxygen make carbon dioxide.

To describe these things in a simple way Berzelius made each of his symbols stand not only for a specific element but also for its relative weight as compared to the weight of other elements—its "atomic weight." To show the composition of a compound he simply put together the symbols for the elements into a "formula"—CO, HCl, FeS, and so on."CO" then not only meant that one atom of carbon and one atom of oxygen combine to make one molecule of carbon monoxide, but also that 12 weight units of carbon (12 being the atomic weight of carbon) combine with 16 weight units of oxygen (16 being the atomic weight of oxygen) to form 28 weight units of the compound carbon monoxide.

When a compound contained several atoms of the same element Berzelius indicated this by placing a number in front of the symbol. It was later found necessary to change this to a smaller number, called a "subscript," placed at the lower right of the symbol — H_2O, CO_2.

In recent years it has been necessary to change Dalton's idea of an atom as being the smallest indivisible part of an element. Nowadays we have machines, such as the cyclotron, that can bombard, or "smash" atoms into still smaller parts—neutrons, and electrically charged protons and electrons. According to today's atomic theory protons and neutrons form the nucleus of the atom and electrons whirl around the nucleus with such tremendous speed that they seem to form a "shell" around it.

But even with our new idea of an atom, Dalton's main theory is still useful for explaining chemical reactions, and Berzelius' method is still the simplest "shorthand" method any scientist has ever devised for writing them down.

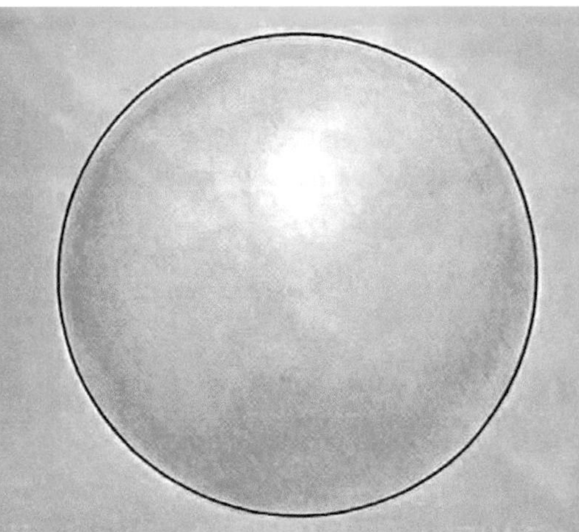

AN ATOM MIGHT LOOK LIKE A BALL SUCH AS THIS IF YOU ENLARGED IT A BILLION TIMES. THE "SHELL" IS NOT SOLID—IT CONSISTS OF ELECTRONS MOVING SO FAST THAT THEY SEEM TO FORM A SOLID SHELL.

IF YOU COULD SLOW DOWN AN ENLARGED CARBON ATOM YOU MIGHT SEE TWO OF ITS ELECTRONS TRAVELING AROUND THE NUCLEUS IN AN "INNER SHELL" AND FOUR MORE WHIRLING AROUND IN AN "OUTER SHELL."

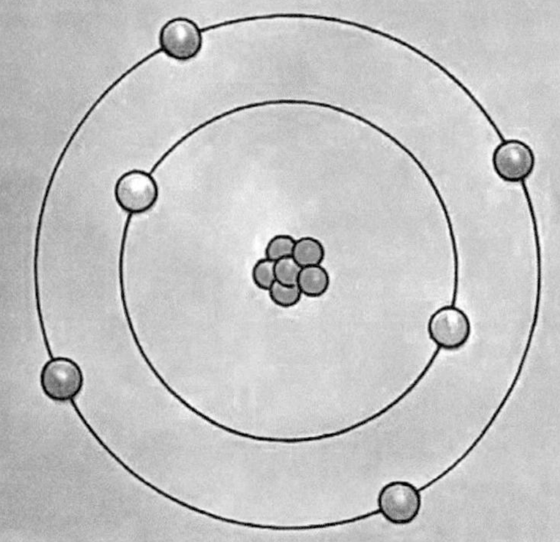

IF YOU COULD HALT AN ENLARGED CARBON ATOM COMPLETELY, IT WOULD LOOK A LOT LIKE OUR SOLAR SYSTEM, WITH A "SUN" (PROTONS AND NEUTRONS) IN THE CENTER AND "PLANETS" (ELECTRONS) AROUND IT.

The Periodic Table of the Elements

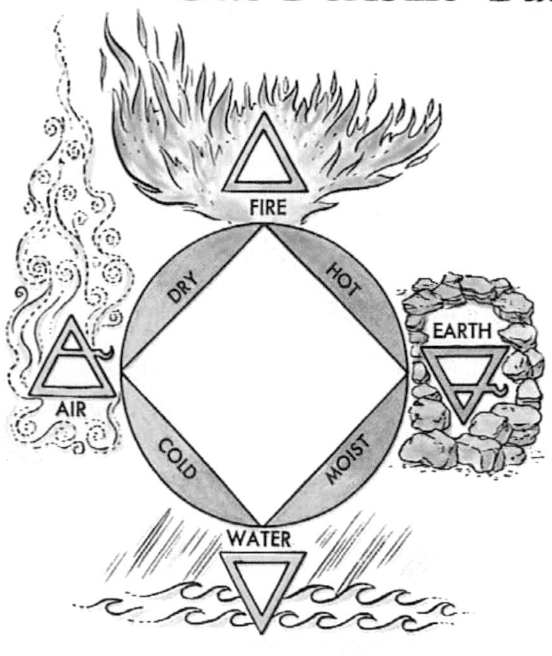

FROM THE earliest times people have tried to explain what "matter" was made of. Most early philosophers agreed that "matter" was made up of what they called "elements." But their idea of an "element" was quite different from what we mean by that word today.

The early Greek philosophers thought the entire universe was composed of only four basic substances: fire, earth, water, and air. This explanation made sense at the time and was not seriously challenged for many centuries.

The old Romans actually knew nine of the substances we call elements today. They called them, of course, by their Latin names (the same we use today in chemical symbols): *carbo* (carbon — C), *sulfur* (S), *aurum* (gold — Au), *argentum* (silver —

FOR MORE THAN A THOUSAND YEARS PHILOSOPHERS INSISTED THAT ALL SUBSTANCES WERE MADE UP OF FOUR ELEMENTS: FIRE THAT WAS DRY AND HOT, EARTH THAT WAS HOT AND MOIST, WATER THAT WAS MOIST AND COLD, AIR THAT WAS COLD AND DRY. WE KNOW BETTER NOW!

hydrogen

1 p

oxygen

8 p

8 n

THE MODERN PICTURE OF AN ATOM HAS A NUCLEUS IN THE CENTER, CONSISTING OF PROTONS (p) AND NEUTRONS (n), WITH ELECTRONS IN RINGS AROUND IT.

THE PERIODIC TABLE OF THE ELEMENTS

	O	I A	II A	III A	IV A	V A	VI A	VII A	
1		H 1 Hydrogen 1.008							
2	He 2 Helium 4.003	Li 3 Lithium 6.940	Be 4 Beryllium 9.013						
3	Ne 10 Neon 20.183	Na 11 Sodium 22.991	Mg 12 Magnesium 24.32						
4	A 18 Argon 39.944	K 19 Potassium 39.1	Ca 20 Calcium 40.08	Sc 21 Scandium 44.96	Ti 22 Titanium 47.9	V 23 Vanadium 50.95	Cr 24 Chromium 52.01	Mn 25 Manganese 54.94	Fe 26 Iron 55.85
5	Kr 36 Krypton 83.8	Rb 37 Rubidium 85.48	Sr 38 Strontium 87.63	Y 39 Yttrium 88.92	Zr 40 Zirconium 91.22	Nb 41 Niobium 92.91	Mo 42 Molybdenum 95.95	Tc 43 Technetium 99	Ru 44 Ruthenium 101.1
6	Xe 54 Xenon 131.3	Cs 55 Cesium 132.91	Ba 56 Barium 137.36	57-71 Lanthanons	Hf 72 Hafnium 178.50	Ta 73 Tantalum 180.95	W 74 Tungsten 183.86	Re 75 Rhenium 186.22	Os 76 Osmium 190.2
7	Rn 86 Radon 222	Fr 87 Francium 223	Ra 88 Radium 226.05	89-103 Actinons					

INERT GASES — ALKALI METALS — ALKALINE EARTH METALS

La 57 Lanthanum 138.92	Ce 58 Cerium 140.13	Pr 59 Praseodymium 140.92	Nd 60 Neodymium 144.27	Pm 61 Promethium 145	Sm 62 Samarium 150.35
Ac 89 Actinium 227	Th 90 Thorium 232.05	Pa 91 Protactinium 231	U 92 Uranium 238.07	Np 93 Neptunium 237	Pu 94 Plutonium 242

ROWS RUNNING FROM LEFT TO RIGHT ARE CALLED PERIODS. COLUMNS RUNNING FROM TOP TO BOTTOM ARE CALLED GROUPS. THE ELEMENTS WITHIN A GROUP HAVE MANY TRAITS IN COMMON.

Ag), *ferrum* (iron — Fe), *cuprum* (copper — Cu), *stannum* (tin — Sn), *plumbum* (lead — Pb), *hydrargyrum* (mercury—Hg).

By 1800, thirty-four elements had been discovered. Within the next ten years, thirteen more had been added and had been given made-up Latin names — among them *natrium* (sodium — Na), *kalium* (potassium — K), and *aluminium* (aluminum — Al). By the beginning of the twentieth century, eighty-four elements were known.

Today the number has reached 102 — the last ten man-made, produced by splitting the atoms of other elements. Within a short time, Element 103 will probably be discovered.

In this table you will find listed the 102 elements that are known today. Each element is described by its chemical symbol, its atomic number, its full name, and its atomic weight.

MANY SCIENTISTS HAD NOTICED THAT IF YOU LINE UP THE ELEMENTS ACCORDING TO ATOMIC WEIGHTS, CERTAIN CHEMICAL TRAITS OCCUR PERIODICALLY. THE RUSSIAN SCIENTIST, DMITRI MENDELEEFF, ON THIS BASIS DISCOVERED THE PERIODIC LAW AND DEVELOPED THE PERIODIC TABLE.

A YOUNG ENGLISH SCIENTIST, HENRY MOSELEY, PERFECTED THE PERIODIC TABLE. HE DISCOVERED THE LAW OF ATOMIC NUMBERS AND ARRANGED THE ELEMENTS ACCORDING TO THE ELECTRIC CHARGE FOUND IN THE NUCLEUS.

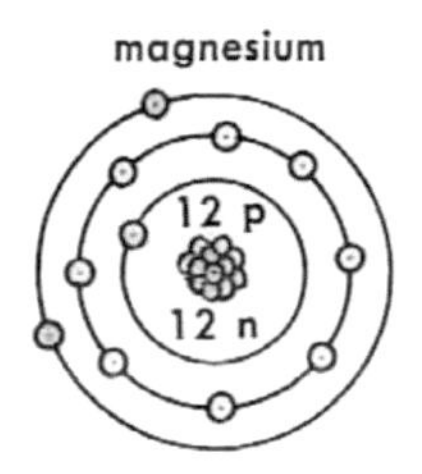

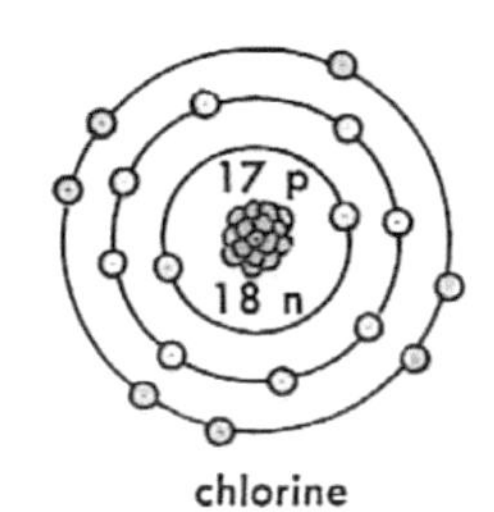

THE NUMBER OF PROTONS IN AN ATOM IS ITS ATOMIC NUMBER. AN ATOM ALWAYS HAS THE SAME NUMBER OF PROTONS AND ELECTRONS. HYDROGEN IS THE SIMPLEST OF ALL ATOMS.

VIII		I B	II B	III B	IV B	V B	VI B	VII B
				B 5 Boron 10.82	C 6 Carbon 12.011	N 7 Nitrogen 14.008	O 8 Oxygen 16	F 9 Fluorine 19
				Al 13 Aluminum 26.98	Si 14 Silicon 28.09	P 15 Phosphorus 30.975	S 16 Sulfur 32.066	Cl 17 Chlorine 35.457
Co 27 Cobalt 58.94	Ni 28 Nickel 58.71	Cu 29 Copper 63.54	Zn 30 Zinc 65.38	Ga 31 Gallium 69.72	Ge 32 Germanium 72.6	As 33 Arsenic 74.91	Se 34 Selenium 78.96	Br 35 Bromine 79.916
Rh 45 Rhodium 102.91	Pd 46 Palladium 106.4	Ag 47 Silver 107.88	Cd 48 Cadmium 112.41	In 49 Indium 114.82	Sn 50 Tin 118.7	Sb 51 Antimony 121.76	Te 52 Tellurium 127.61	I 53 Iodine 126.91
Ir 77 Iridium 192.2	Pt 78 Platinum 195.09	Au 79 Gold 197	Hg 80 Mercury 200.61	Tl 81 Thallium 204.39	Pb 82 Lead 207.21	Bi 83 Bismuth 209	Po 84 Polonium 210	At 85 Astatine 210

HEAVY METALS | METALLOIDS | NON-METALS | RARE EARTH METALS | UNSTABLE ELEMENTS

Eu 63 Europium 152	Gd 64 Gadolinium 157.26	Tb 65 Terbium 158.93	Dy 66 Dysprosium 162.51	Ho 67 Holmium 164.94	Er 68 Erbium 167.27	Tm 69 Thulium 168.94	Yb 70 Ytterbium 173.04	Lu 71 Lutetium 174.99
Am 95 Americium 243	Cm 96 Curium 247	Bk 97 Berkelium 249	Cf 98 Californium 249	E 99 Einsteinium 254	Fm 100 Fermium 255	Mv 101 Mendelevium 256	No 102 Nobelium 251	? 103

The Mysteries of Solutions

SVANTE ARRHENIUS DEVELOPED THEORY TO EXPLAIN HOW SOLUTIONS CONDUCT ELECTRICITY.

FROM THE earliest days, scientists experimenting with chemistry have worked with solutions. The liquid they used for making a solution (usually water) they called the "solvent." The chemical dissolved was the "solute."

When chemists began to use electricity as one of their tools, they discovered that different solutions behaved in different ways. The solution in water of a great number of chemicals — sugar among them — did not let electricity pass through. They were "non-conductors." Some chemicals, on the other hand, conducted electricity very easily. They were good conductors — "electrolytes."

In 1874 a Swedish scientist named Svante Arrhenius developed a theory to help explain the mysterious behavior of solutions. He was only 25 years old at the time.

His idea was that when a chemical that conducts electricity is dissolved in water, each molecule is broken up — "dissociated" — into electrically charged atoms or groups of atoms. These atoms or groups of atoms Arrhenius called "ions" from a Greek word that means "to wander." His new theory came to be called "Arrhenius' theory of ionization."

When table salt (sodium chloride, $NaCl$), for instance, is dissolved in water, it ionizes into positively charged sodium ions (Na^+) and negatively charged chlorine ions (Cl^-). These ions "wander" about in all directions until an electric current is applied to the solution. When that happens, the negative ions rush to the positive pole, the positive ions to the negative pole. It is the ions that conduct the current through the solution.

The reason that non-conductors do not conduct electricity is that they do not dissociate into ions.

Arrhenius' theory of ionization helped explain a great number of things that have puzzled chemists. His theory has been modified somewhat over the years but in most respects holds true today.

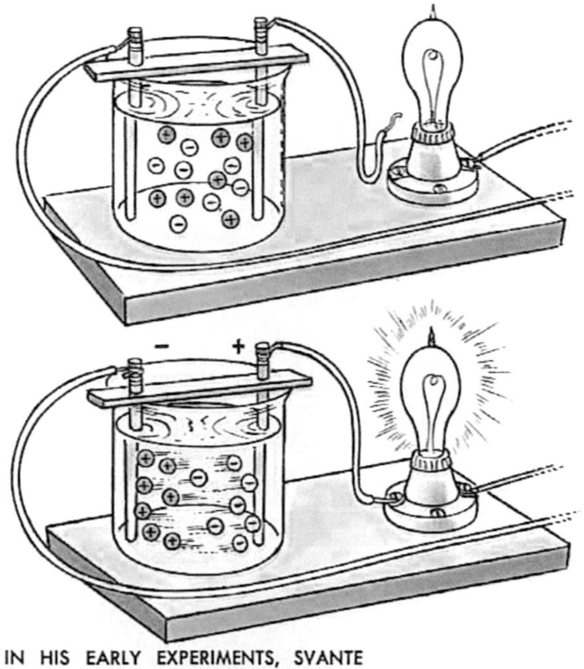

IN HIS EARLY EXPERIMENTS, SVANTE ARRHENIUS USED A SIMPLE SET-UP. YOU CAN EASILY REPEAT SOME OF HIS EXPERIMENTS IN YOUR OWN LAB, USING FLASHLIGHT BATTERIES.

TESTING CONDUCTIVITY OF SOLUTIONS

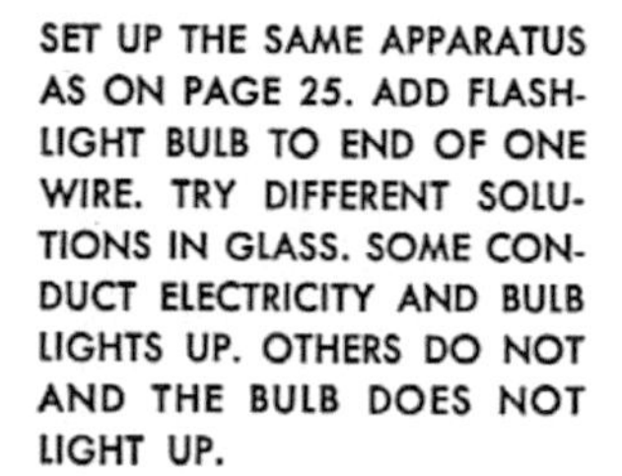

SET UP THE SAME APPARATUS AS ON PAGE 25. ADD FLASHLIGHT BULB TO END OF ONE WIRE. TRY DIFFERENT SOLUTIONS IN GLASS. SOME CONDUCT ELECTRICITY AND BULB LIGHTS UP. OTHERS DO NOT AND THE BULB DOES NOT LIGHT UP.

SATURATED SOLUTIONS

A SATURATED SOLUTION IS ONE IN WHICH NO MORE OF THE CHEMICAL WILL GO IN SOLUTION AT THAT PARTICULAR TEMPERATURE.

1 POUR 20 ml WATER OF ROOM TEMPERATURE INTO A CUSTARD CUP. ADD 6 g SALTPETER (POTASSIUM NITRATE). STIR. ALL THE SALTPETER DISSOLVES.

2 ADD 3 g MORE SALTPETER. STIR. SOME OF THE ADDED SALTPETER DOES NOT DISSOLVE. CLEAR LIQUID IS SATURATED AT ROOM TEMPERATURE. (AT 20°C., 6.3 g KNO_3 MAKES SATURATED SOLUTION IN 20 ml WATER.)

3 PLACE CUSTARD CUP OVER ALCOHOL BURNER. ADD 10 g MORE SALTPETER. SOON ALL SALTPETER IS DISSOLVED. AT HIGHER TEMPERATURES IT TAKES MORE SOLUTE TO MAKE A SATURATED SOLUTION. (AT BOILING, 20 ml H_2O DISSOLVES 49 g SALTPETER.)

4 TAKE SOLUTION OFF FIRE. AS IT COOLS, MUCH OF THE SALTPETER COMES OUT AS CRYSTALS BY SLOW CRYSTALLIZATION. LIQUID IS AGAIN A SOLUTION SATURATED AT ROOM TEMPERATURE.

3

$NaNO_3$

4

$Na_2S_2O_3 \cdot 5H_2O$

BEHAVIOR OF SOLUTIONS

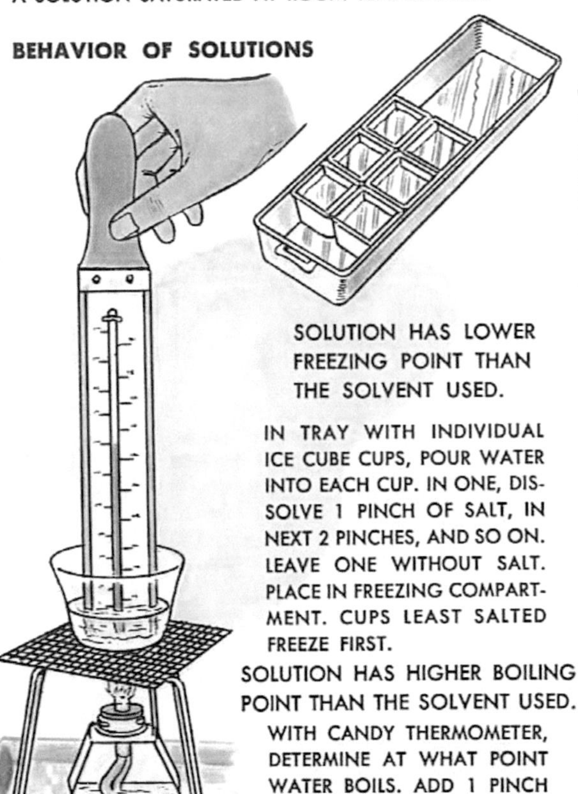

SOLUTION HAS LOWER FREEZING POINT THAN THE SOLVENT USED.

IN TRAY WITH INDIVIDUAL ICE CUBE CUPS, POUR WATER INTO EACH CUP. IN ONE, DISSOLVE 1 PINCH OF SALT, IN NEXT 2 PINCHES, AND SO ON. LEAVE ONE WITHOUT SALT. PLACE IN FREEZING COMPARTMENT. CUPS LEAST SALTED FREEZE FIRST.

SOLUTION HAS HIGHER BOILING POINT THAN THE SOLVENT USED.

WITH CANDY THERMOMETER, DETERMINE AT WHAT POINT WATER BOILS. ADD 1 PINCH OF SALT. WHAT IS BOILING POINT NOW? ADD MORE SALT. READ AGAIN.

CRYSTALLIZATION

YOU CAN FOLLOW CRYSTALLIZATION OF $MgSO_4$. IN TEST TUBE, HEAT MIXTURE OF 5 ml WATER AND 1 TEASPOON EPSOM SALT UNTIL SALT DISSOLVES. POUR HOT SOLUTION OVER PANE OF GLASS CLEANED WITH DETERGENT. CRYSTALS MAKE NEEDLE-LIKE NETWORK.

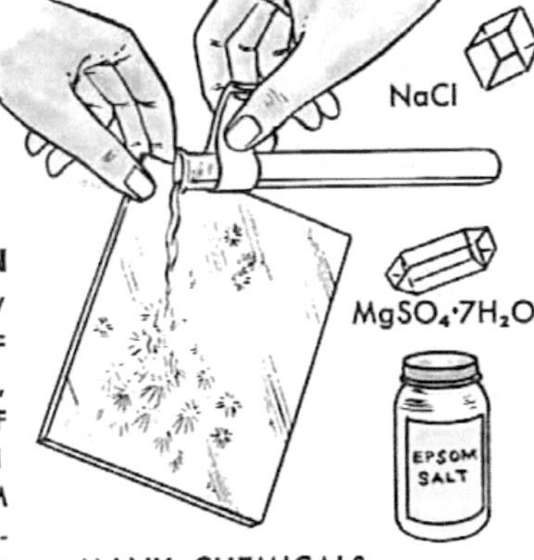

MANY CHEMICALS FORM CRYSTALS OF DISTINCT SHAPES.

$FeSO_4 \cdot 7H_2O$

MAKING SOLUTIONS

MAKE 50 ml GRADUATE FIRST: MEASURE 50 ml WATER INTO A NARROW JAR, USING 10 ml TEST TUBE GRADUATE SHOWN ON PAGE 15. MAKE A MARK AT 50 ml LEVEL.

10% (10 PER CENT) SOLUTION: MEASURE 40 ml WATER INTO A CUSTARD CUP. ADD 5 g OF THE CHEMICAL. STIR. (TO MAKE IT DISSOLVE QUICKER, YOU MAY WANT TO HEAT THE WATER SLIGHTLY.) POUR SOLUTION INTO 50 ml GRADUATE. ADD WATER TO THE 50 ml MARK.

2% SOLUTION: MEASURE 40 ml WATER INTO CUSTARD CUP. ADD 1 g OF THE CHEMICAL. STIR TO DISSOLVE. POUR INTO 50 ml GRADUATE. ADD WATER TO 50 ml.

HOW DO YOU KNOW AN ACID?

1. ACIDS TASTE SOUR.

ADD 5 ml HYDROCHLORIC ACID TO 15 ml WATER. DROP 5 DROPS OF MIXTURE IN GLASS OF WATER. DIP FINGER IN THIS HIGHLY DILUTED ACID. TASTE DROP ON FINGER TIP.

HYDROCHLORIC ACID

2. ACIDS ACT WITH INDICATORS.

PLACE DROP OF DILUTED HYDROCHLORIC ACID ON STRIP OF BLUE LITMUS PAPER. THE COLOR CHANGES TO RED.

BLUE LITMUS

3. ACIDS ACT WITH METALS.

HYDROCHLORIC ACID

PLACE A STRIP OF ZINC IN A TEST TUBE. POUR A FEW ml HYDROCHLORIC ACID ON IT. ZINC DISSOLVES, SETTING THE HYDROGEN OF ACID FREE.

4. ACIDS NEUTRALIZE BASES.

1 COLOR 2 ml LYE SOLUTION WITH A DROP OF PHENOLPHTHALEIN SOLUTION.

2 POUR INTO 5 ml HYDROCHLORIC ACID. THE PINK COLOR DISAPPEARS.

Working With Acids

ACIDS have many traits in common. They taste sour. They change the color of certain plant substances—which are called "indicators." They contain hydrogen (H) that can be replaced by a metal. They neutralize bases.

But what *is* an acid? Earlier, the "acidic" traits were used to define an acid. But with the modern understanding of the atom, a different definition is used. You will remember that the nucleus of an atom contains positively charged protons. Acids in solution liberate protons as ions (H^+). And so we say that an acid is a substance that will give up — or "donate" — protons to another substance. Acids are "proton donors." The foremost acids used in industry are sulfuric acid (H_2SO_4), nitric acid (HNO_3), and hydrochloric acid (HCl).

The first two — sulfuric acid and nitric acid — should NEVER be used in the home lab. They are much too DANGEROUS. They destroy the skin and might blind you if you got them in the eyes. (Wherever a chemical experiment would ordinarily call for sulfuric acid, this book uses sodium acid sulfate — $NaHSO_4$, sodium bisulfate, "Sani-Flush"; wherever

(CONTINUED ON PAGE 44)

HOME-MADE INDICATORS

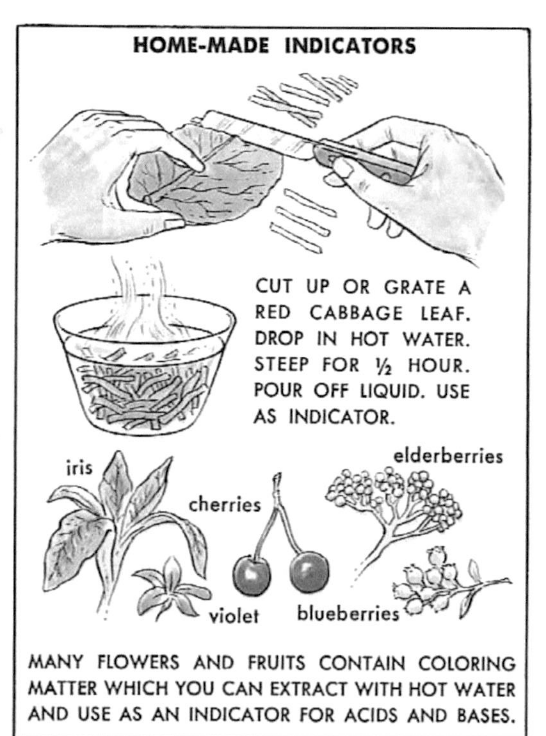

MANY FLOWERS AND FRUITS CONTAIN COLORING MATTER WHICH YOU CAN EXTRACT WITH HOT WATER AND USE AS AN INDICATOR FOR ACIDS AND BASES.

Working With Bases

BASES taste brackish. They change the color of "indicators." They contain a combination of oxygen and hydrogen atoms called "hydroxyl" (OH). They neutralize acids.

But what *is* a base? When a base is dissolved in water it liberates negatively charged hydroxyl ions (OH^-). When a base is neutralized, these ions take on — or "accept"— positively charged protons from another substance. A base is a substance that will accept and combine with protons from another substance. Bases are "proton acceptors." The most important bases are sodium hydroxide ("lye," NaOH), ammonium hydroxide ("ammonia," NH_4OH), and calcium hydroxide ("slaked lime," $Ca(OH)_2$).

The first of these — sodium hydroxide — is used in many households to clean sluggish drains and to keep sinks from stopping up ("Drano"). USE IT WITH GREAT CARE in your experiments. Do not touch lye flakes with your fingers and do not get the solution on your skin — it dissolves the natural oil. It is particularly dangerous to get lye in your eyes. If you get lye on you, dilute it quickly with LOTS OF WATER.

(CONTINUED ON PAGE 45)

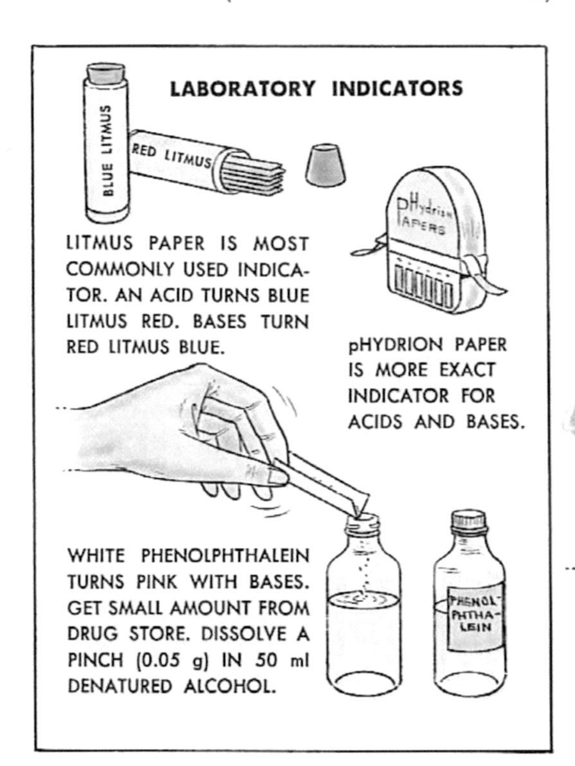

LITMUS PAPER IS MOST COMMONLY USED INDICATOR. AN ACID TURNS BLUE LITMUS RED. BASES TURN RED LITMUS BLUE.

pHYDRION PAPER IS MORE EXACT INDICATOR FOR ACIDS AND BASES.

WHITE PHENOLPHTHALEIN TURNS PINK WITH BASES. GET SMALL AMOUNT FROM DRUG STORE. DISSOLVE A PINCH (0.05 g) IN 50 ml DENATURED ALCOHOL.

HOW DO YOU KNOW A BASE?

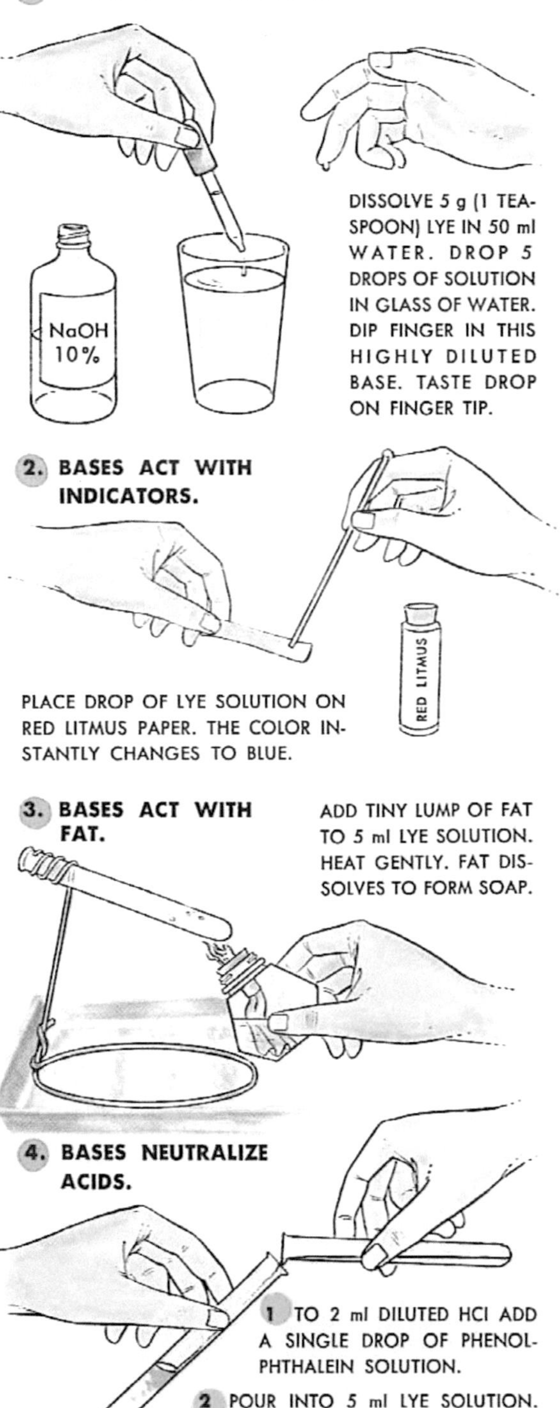

1. BASES TASTE BRACKISH.

DISSOLVE 5 g (1 TEASPOON) LYE IN 50 ml WATER. DROP 5 DROPS OF SOLUTION IN GLASS OF WATER. DIP FINGER IN THIS HIGHLY DILUTED BASE. TASTE DROP ON FINGER TIP.

2. BASES ACT WITH INDICATORS.

PLACE DROP OF LYE SOLUTION ON RED LITMUS PAPER. THE COLOR INSTANTLY CHANGES TO BLUE.

3. BASES ACT WITH FAT.

ADD TINY LUMP OF FAT TO 5 ml LYE SOLUTION. HEAT GENTLY. FAT DISSOLVES TO FORM SOAP.

4. BASES NEUTRALIZE ACIDS.

1 TO 2 ml DILUTED HCl ADD A SINGLE DROP OF PHENOLPHTHALEIN SOLUTION.

2 POUR INTO 5 ml LYE SOLUTION. THE MIXTURE TURNS A BRILLIANT PINK.

HOUSEHOLD ITEMS CONTAINING ACIDS

LIME
ORANGE
TOMATO
VINEGAR
GINGER ALE
Tea
SOUR CREAM
ASPIRIN

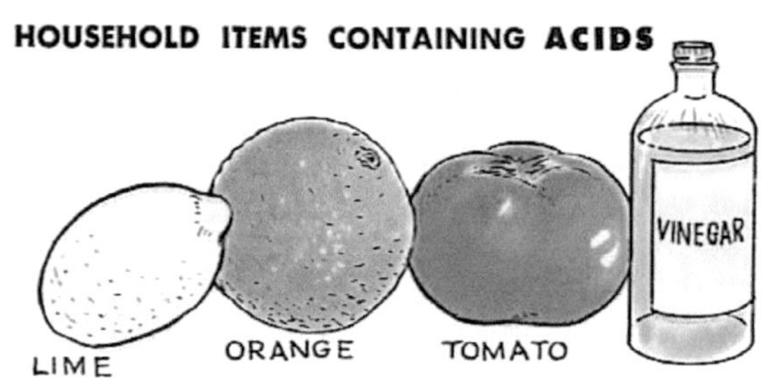

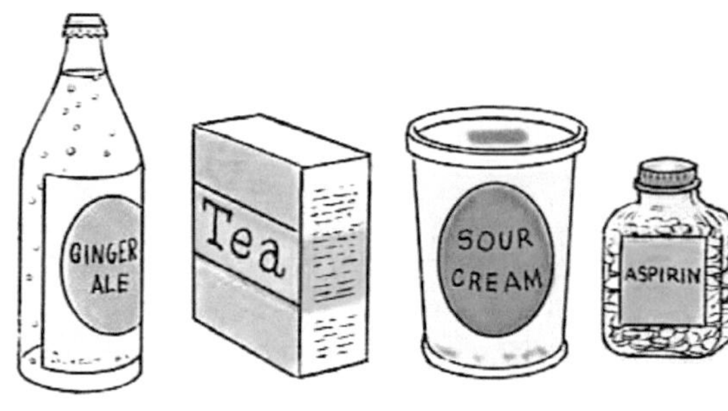

Acids—Continued

nitric acid would be called for, this book produces it in a mixture of a nitrate, KNO_3, and sodium bisulfate.)

Hydrochloric acid is used in many households under the name of "muriatic acid." Whenever you use hydrochloric acid in an experiment; USE IT WITH GREAT CARE. If any of it gets on you, dilute it quickly with LOTS OF WATER. Or neutralize it with bicarbonate of soda (but not if in the eyes).

ACID FROM NON-METALLIC OXIDE

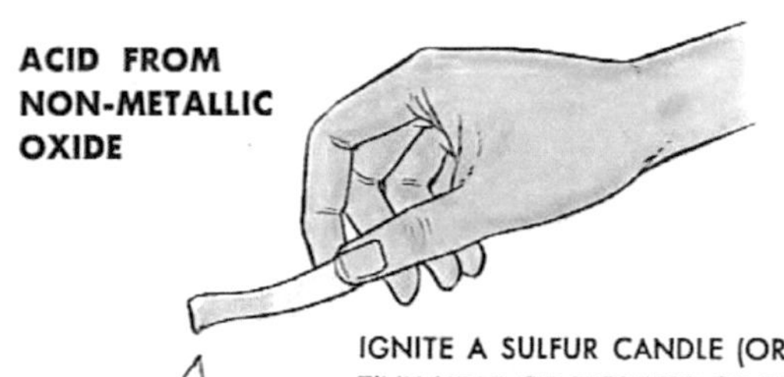

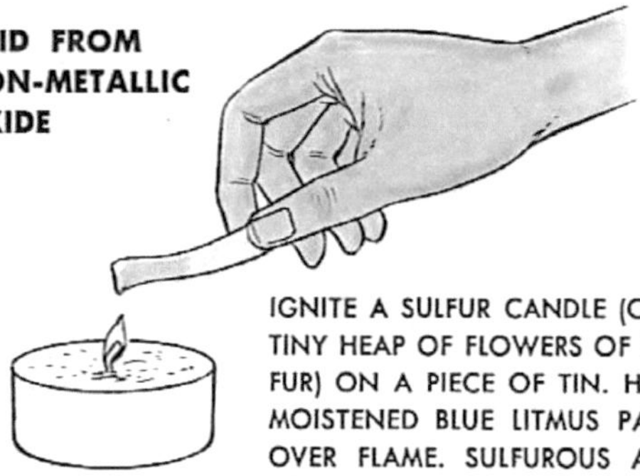

IGNITE A SULFUR CANDLE (OR A TINY HEAP OF FLOWERS OF SULFUR) ON A PIECE OF TIN. HOLD MOISTENED BLUE LITMUS PAPER OVER FLAME. SULFUROUS ACID FORMED TURNS IT RED.

ACID FROM A SALT

SALT
A
blue litmus turns red from acid.
B

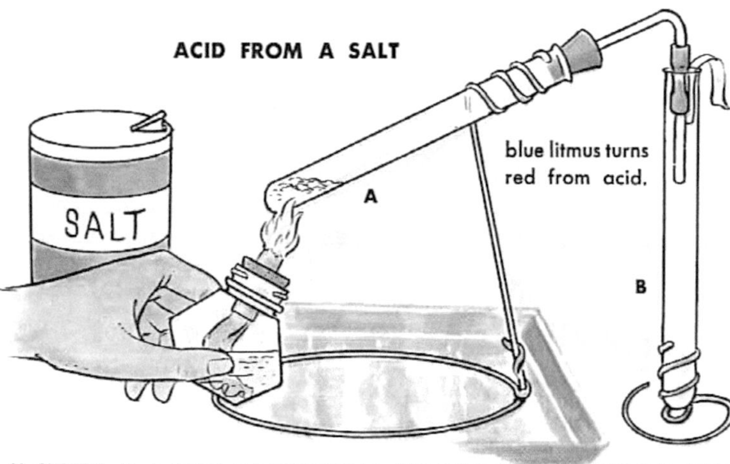

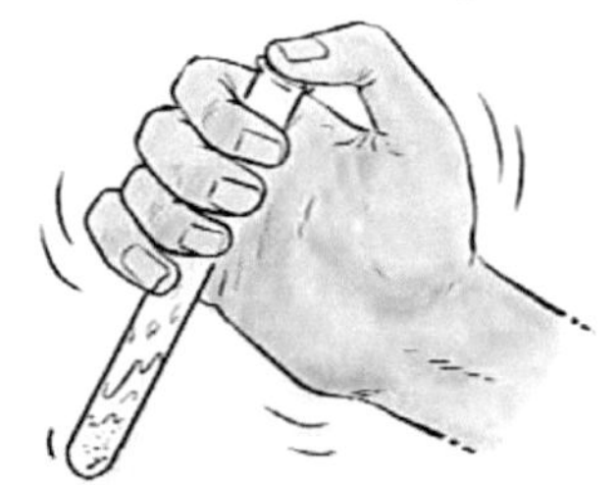

SET UP APPARATUS AS SHOWN. INTO TEST TUBE **A** DROP MIXTURE OF ¼ TEASPOON TABLE SALT AND ½ TEASPOON SODIUM BISULFATE. HEAT. HYDROGEN CHLORIDE PRODUCED TURNS MOISTENED BLUE LITMUS RED. ADD 2 ml WATER TO TEST TUBE **B**. SHAKE. RESULT IS WEAK HYDROCHLORIC ACID.

pH SYSTEM IS A WAY OF DESCRIBING THE RELATIVE ACIDITY OR ALKALINITY OF A SOLUTION. PURE WATER IS NEUTRAL WITH pH7. THE LOWER THE NUMBER BELOW 7, THE MORE ACID THE SOLUTION. THE HIGHER

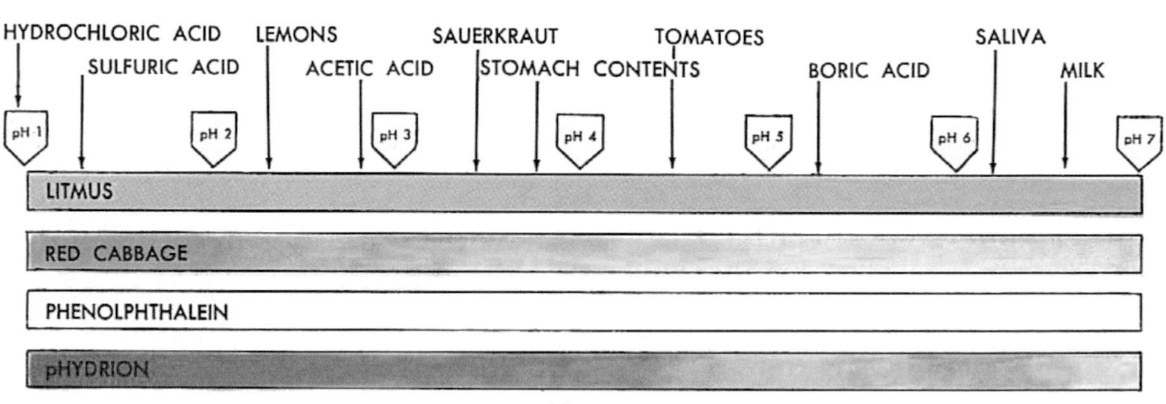

HYDROCHLORIC ACID
SULFURIC ACID
LEMONS
ACETIC ACID
SAUERKRAUT
STOMACH CONTENTS
TOMATOES
BORIC ACID
SALIVA
MILK
pH 1
pH 2
pH 3
pH 4
pH 5
pH 6
pH 7
LITMUS
RED CABBAGE
PHENOLPHTHALEIN
pHYDRION

HOUSEHOLD ITEMS CONTAINING BASES

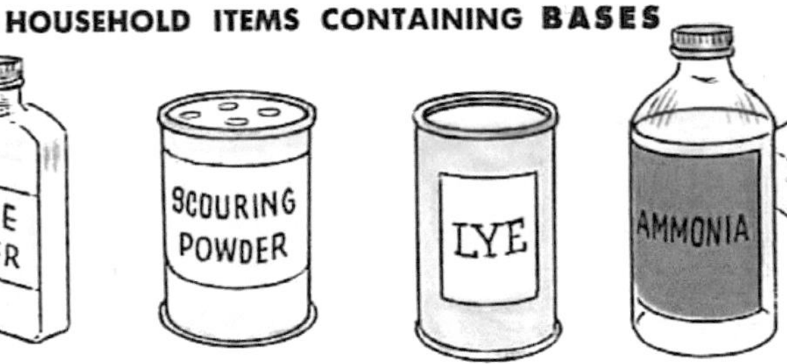

Bases—Continued

You can also neutralize it with vinegar (but not if in the eyes).

Ammonia is a common household cleaning liquid. Ammonia should also be handled with care and should be washed off quickly if you get it on you. Also watch your nose when you work with ammonia. It has a very strong smell.

Calcium hydroxide is a white powder. You will use it in a great number of experiments.

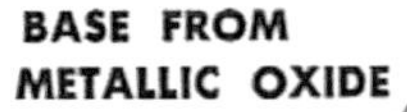

BASE FROM METALLIC OXIDE

PLACE A LUMP OF LIME (QUICKLIME, CALCIUM OXIDE) IN A CUSTARD CUP. ADD AS MUCH LUKEWARM WATER AS IT WILL ABSORB. LIME HEATS UP, GIVES OFF STEAM, CRUMBLES INTO POWDER OF SLAKED LIME (CALCIUM HYDROXIDE).

BASE FROM A SALT

IN A CUSTARD CUP, DISSOLVE 1 TEASPOON SAL SODA (WASHING SODA, SODIUM CARBONATE) IN 50 ml WATER. HEAT SLIGHTLY. ADD SLAKED LIME MIXED WITH WATER. STIR. CHEMICAL REACTION PRODUCES SODIUM HYDROXIDE AND CALCIUM CARBONATE. FILTER. CLEAR LIQUID CONTAINS THE SODIUM HYDROXIDE (LYE). THE CALCIUM CARBONATE IS HELD BACK BY THE FILTER.

red litmus turns blue from base.

SAL SODA CONCENTRATED WASHING SODA

THE NUMBER ABOVE 7, THE MORE ALKALINE THE SOLUTION. WHEN YOU KNOW AT WHAT pH AN INDICATOR CHANGES COLOR, YOU CAN DETERMINE THE ACIDITY OR ALKALINITY OF THE SOLUTION YOU ARE TESTING.

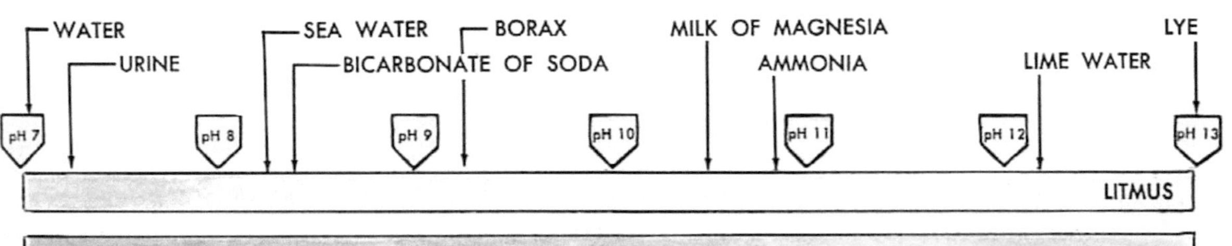

Salts—Chemicals of Many Uses

NEUTRALIZATION IS USED EXTENSIVELY IN CHEMICAL ANALYSIS IN A TECHNIQUE CALLED **TITRATION.**

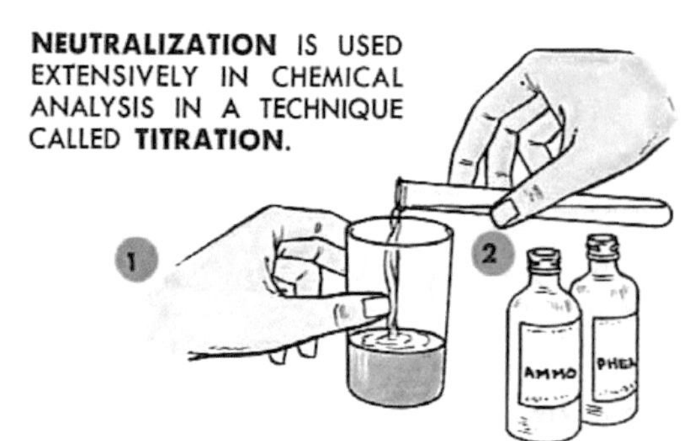

TO DETERMINE THE UNKNOWN STRENGTH OF A BASE, THE CHEMIST DROPS INTO IT FROM A LONG TUBE—A BURETTE—AS MUCH ACID OF KNOWN STRENGTH AS IS NECESSARY TO NEUTRALIZE IT. BY CHECKING ACID USED HE FIGURES STRENGTH OF BASE.

1 FOR A TRY AT TITRATION, MIX A FEW ml OF HOUSEHOLD AMMONIA WITH 40 ml WATER. ADD A DROP OF PHENOLPHTHALEIN. THIS WILL COLOR THE MIXTURE A DEEP PINK.

2 POUR 10 ml DILUTED HYDROCHLORIC ACID INTO MEASURING TUBE. POUR SOME OF THIS ACID INTO THE AMMONIA UNTIL COLOR HAS ALMOST VANISHED.

3 PICK UP A FEW ml OF THE MEASURED ACID IN AN EYE DROPPER (PIPETTE). DROP ACID SLOWLY INTO THE AMMONIA MIXTURE UNTIL COLOR IS COMPLETELY GONE. RETURN ACID NOT USED TO MEASURING TUBE. YOU NOW KNOW HOW MANY ml ACID YOU HAD TO USE TO NEUTRALIZE THE AMMONIA.

WHAT HAPPENS when you neutralize an acid with a base or a base with an acid? The hydrogen atoms (H^+ ions) of the acid combine with the hydroxyl groups (OH^- ions) of the base to form water, and the metal atoms of the base combine with what remains of the acid to form a salt. Or simply:

BASE plus ACID turns into
WATER plus SALT

This, for example, is what happens when you neutralize sodium hydroxide with hydrochloric acid:

$$NaOH + HCl \rightarrow HOH + NaCl$$

The result is water and sodium chloride — ordinary table salt which has given its name to other substances of a similar nature.

Of all the salts used in industry, table salt ($NaCl$) and washing soda (Na_2CO_3) are of greatest importance. Numerous other chemicals are produced from them. Our way of life would be completely disrupted if our country's industry did not have enough of these two salts.

Many other salts are necessary for our well-being. You'll probably find at least half a dozen different salts used daily in your home — in cooking and baking, in gardening, for cleaning.

In your chemical experiments you'll be working with two classes of salts: *normal salts* (such as $NaCl$, Na_2CO_3, KI) which contain no free hydrogen or hydroxyl ions, and *acid salts* (such as $NaHSO_4$, $NaHCO_3$) which contain replaceable hydrogen.

Some of these salts dissolve easily in water — all the nitrates (salts of nitric acid) and most of the chlorides (salts of hydrochloric acid). Many salts, on the other hand, are insoluble — most of the carbonates (salts of carbonic acid) and most sulfides (salts of hydrosulfuric acid).

HOW THE NAMES OF SALTS ARE MADE UP

THE ACID	FORMULA AND NAME OF ACID		FORMULA AND NAME OF SALT	
SULFURIC ACID	H_2SO_4	HYDROGEN SULFATE	Na_2SO_4	SODIUM SULFATE
NITRIC ACID	HNO_3	HYDROGEN NITRATE	$NaNO_3$	SODIUM NITRATE
CARBONIC ACID	H_2CO_3	HYDROGEN CARBONATE	Na_2CO_3	SODIUM CARBONATE
ACETIC ACID	$HC_2H_3O_2$	HYDROGEN ACETATE	$NaC_2H_3O_2$	SODIUM ACETATE
HYDROCHLORIC ACID	HCl	HYDROGEN CHLORIDE	$NaCl$	SODIUM CHLORIDE
HYDROSULFURIC ACID	H_2S	HYDROGEN SULFIDE	Na_2S	SODIUM SULFIDE
SULFUROUS ACID	H_2SO_3	HYDROGEN SULFITE	Na_2SO_3	SODIUM SULFITE
NITROUS ACID	HNO_2	HYDROGEN NITRITE	$NaNO_2$	SODIUM NITRITE
CHLOROUS ACID	$HClO_2$	HYDROGEN CHLORITE	$NaClO_2$	SODIUM CHLORITE

REMEMBER: **-IC** ACIDS FORM **-ATE** SALTS;
HYDRO- -IC ACIDS FORM **-IDE** SALTS; **-OUS** ACIDS FORM **-ITE** SALTS

HOUSEHOLD ITEMS CONTAINING SALTS

DIFFERENT WAYS OF PRODUCING SALTS

SALT FROM METAL AND ACID

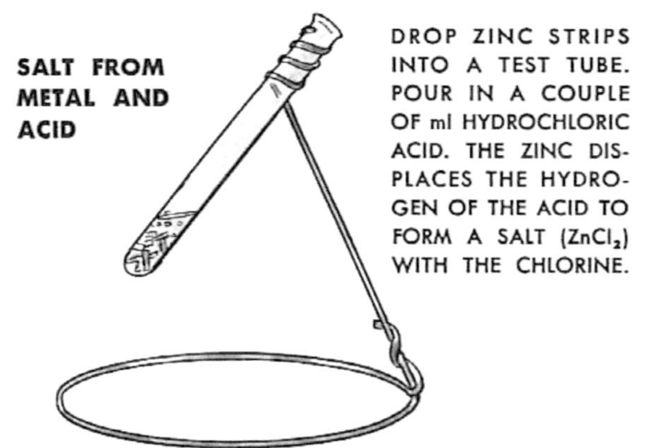

DROP ZINC STRIPS INTO A TEST TUBE. POUR IN A COUPLE OF ml HYDROCHLORIC ACID. THE ZINC DISPLACES THE HYDROGEN OF THE ACID TO FORM A SALT ($ZnCl_2$) WITH THE CHLORINE.

SALT FROM METAL OXIDE AND ACID

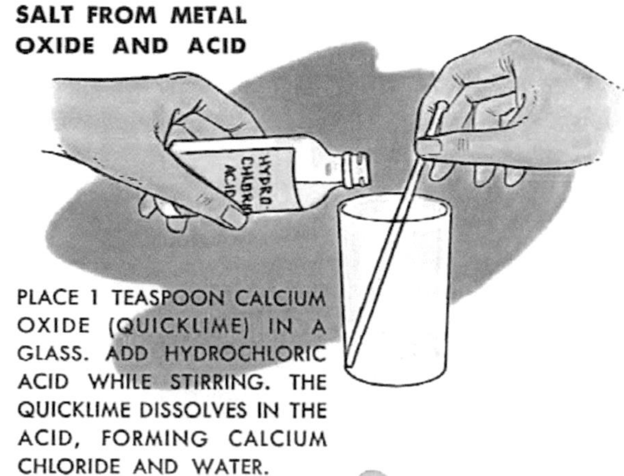

PLACE 1 TEASPOON CALCIUM OXIDE (QUICKLIME) IN A GLASS. ADD HYDROCHLORIC ACID WHILE STIRRING. THE QUICKLIME DISSOLVES IN THE ACID, FORMING CALCIUM CHLORIDE AND WATER.

SALT FROM ANOTHER SALT AND ACID

DROP PIECES OF CHALK, MARBLE, OR OYSTER SHELLS (ALL OF THEM CALCIUM CARBONATES) IN A FEW ml HYDROCHLORIC ACID. RESULT IS CALCIUM CHLORIDE AND CARBONIC ACID (WHICH BREAKS UP INTO CARBON DIOXIDE AND WATER).

TWO SALTS FROM TWO OTHER SALTS

1 DISSOLVE 5 g EPSOM SALTS (MAGNESIUM SULFATE) IN 20 ml. WATER. BRING TO BOIL.

2 DISSOLVE 5 g SODA IN 20 ml WARM WATER. POUR INTO HOT EPSOM SALT SOLUTION.

3 FILTER THE MILKY MIXTURE. THE FILTRATE CONTAINS SODIUM SULFATE. MAGNESIUM CARBONATE IS RETAINED BY FILTER.

THERE ARE many ways of producing a salt in addition to neutralization.

When you made iron sulfide directly from the two elements iron and sulfur, you produced a salt:

$$Fe + S \rightarrow FeS$$

When you caused zinc metal to react with hydrochloric acid, you made a salt:

$$Zn + 2HCl \rightarrow ZnCl_2 + H_2$$

When you made sodium hydroxide, you used a base and a salt to form a new base and a new salt:

$$Ca(OH)_2 + Na_2CO_3 \rightarrow 2NaOH + CaCO_3$$

A salt and an acid often form another salt and another acid:

$$CaCO_3 + 2HCl \rightarrow CaCl_2 + H_2CO_3 \quad (H_2O + CO_2)$$

Two soluble salts may also form two other salts—one of them insoluble:

$$Na_2CO_3 + MgSO_4 \rightarrow MgCO_3 + Na_2SO_4$$

Iodine—Violet or Brown?

IODINE Element 53. At. wt. 126.91. Gray-black crystals of a peculiar odor. Sublimes with violet color. Combines directly with metals and nonmetals. It has a density of 4.9.

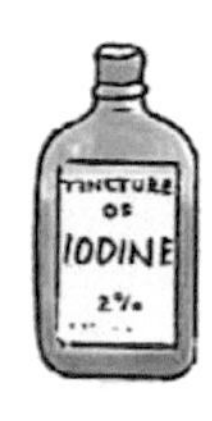

IODINE IS an interesting element to experiment with. It is easily driven out of its compounds as beautiful, violet fumes that turn into grayish-black, metallic-looking crystals on cooling. These crystals can be further purified by turning them into vapor again, and again cooling them into crystal form. This process is called "sublimation."

You are probably familiar with the 2% alcoholic solution of iodine known as "tincture of iodine." It is found in almost every home medicine cabinet and is used as a disinfectant for wounds. Iodine has many other uses — in photography and in the preparation of various medicines and dyes.

Iodine has the bad habit of staining practically everything with which it comes in contact with a brown stain that won't come off in washing. That's why it is advisable to have sodium thiosulfate — photographer's fixing salt, "hypo"— around when you work with iodine. Hypo in solution forms a colorless compound with iodine.

MAKING IODINE

Be careful not to breathe fumes.

1 IN A PYREX CUSTARD CUP MIX TOGETHER 2 g POTASSIUM IODIDE, 2 g MANGANESE DIOXIDE, 4 g SODIUM BISULFATE. HEAT MIXTURE GENTLY. SOON VIOLET FUMES EMERGE.

2 DROP HALF A DOZEN ICE CUBES INTO A JAR. ADD A LITTLE WATER. PLACE JAR AS A LID ON TOP OF CUSTARD CUP. THE VIOLET FUMES SETTLE ON BOTTOM OF JAR AS GRAYISH-BLACK, SHINY IODINE CRYSTALS.

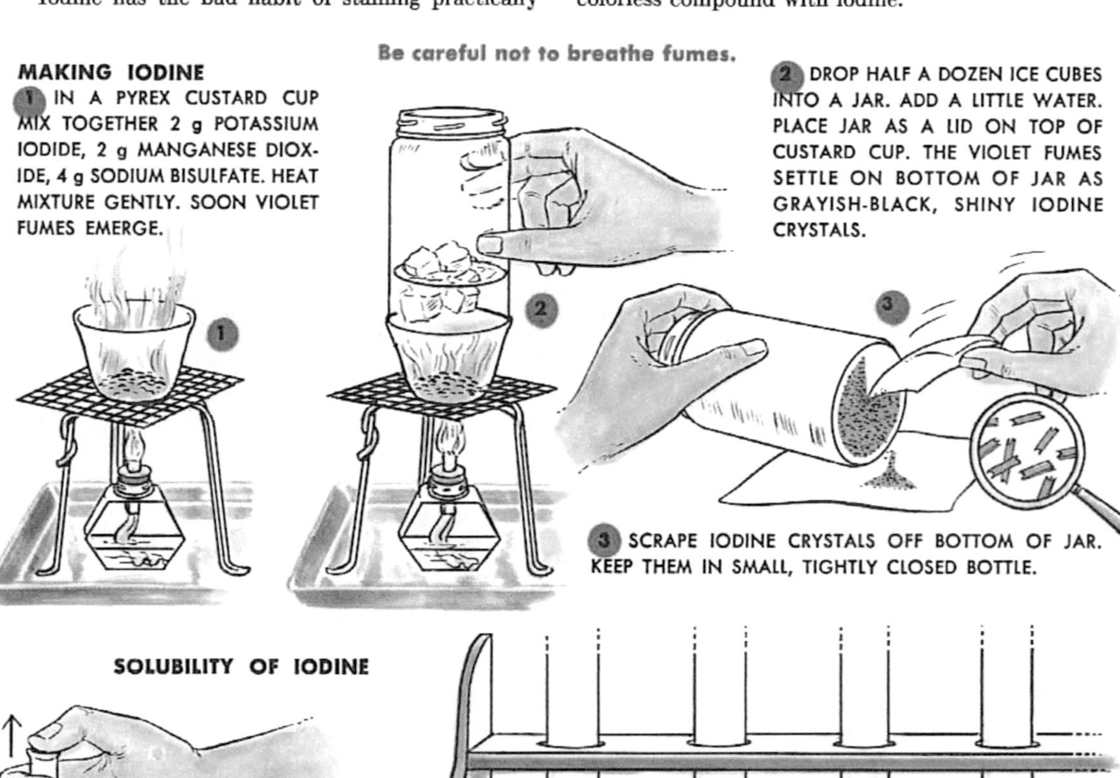

3 SCRAPE IODINE CRYSTALS OFF BOTTOM OF JAR. KEEP THEM IN SMALL, TIGHTLY CLOSED BOTTLE.

SOLUBILITY OF IODINE

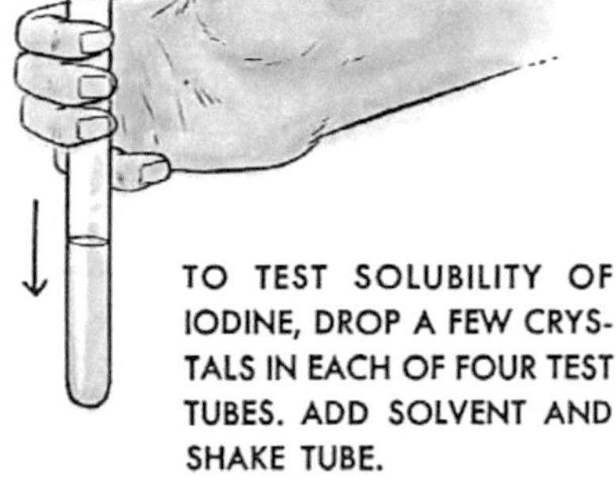

TO TEST SOLUBILITY OF IODINE, DROP A FEW CRYSTALS IN EACH OF FOUR TEST TUBES. ADD SOLVENT AND SHAKE TUBE.

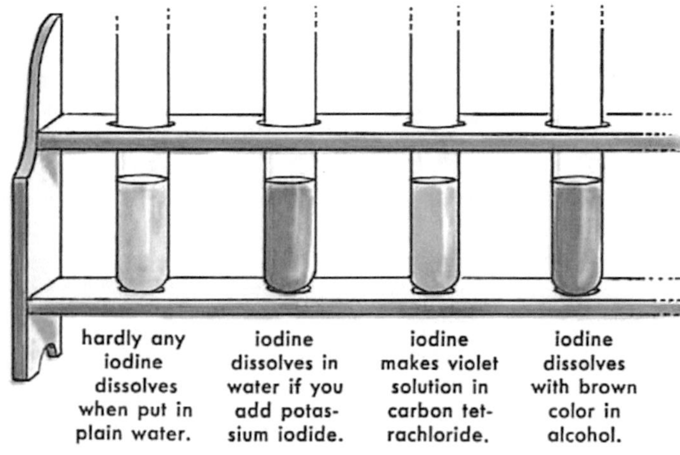

hardly any iodine dissolves when put in plain water. | iodine dissolves in water if you add potassium iodide. | iodine makes violet solution in carbon tetrachloride. | iodine dissolves with brown color in alcohol.

IODINE FREED BY CHLORINE

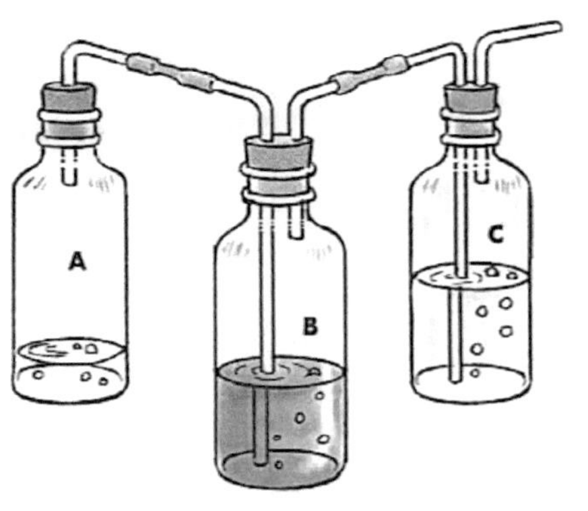

SET UP APPARATUS AS DESCRIBED ON PAGE 35 WITH THIS EXCEPTION: IN BOTTLE **B**, MAKE SOLUTION OF ½ g POTASSIUM IODIDE IN 40 ml WATER. AS CHLORINE BUBBLES THROUGH THIS SOLUTION IT TURNS BROWN FROM THE FREED IODINE. WITH MORE CHLORINE IT CLEARS AGAIN WHEN COLORLESS IODIC ACID FORMS.

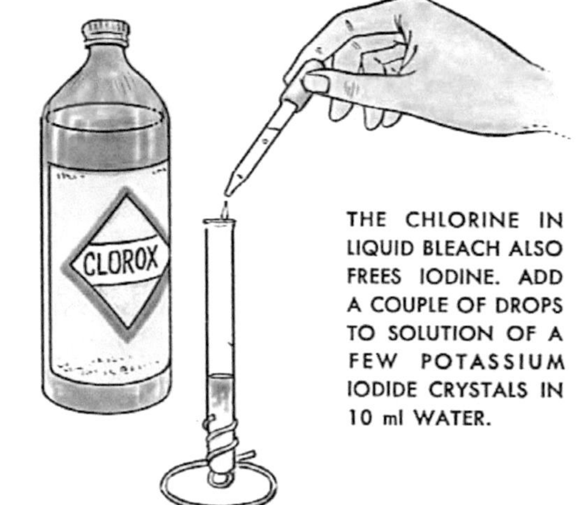

THE CHLORINE IN LIQUID BLEACH ALSO FREES IODINE. ADD A COUPLE OF DROPS TO SOLUTION OF A FEW POTASSIUM IODIDE CRYSTALS IN 10 ml WATER.

IODINE BY OXIDATION

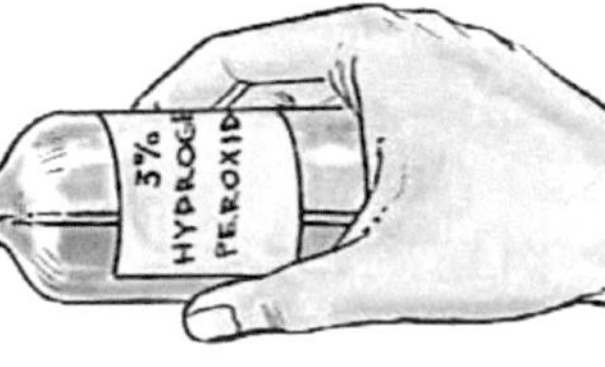

DISSOLVE A FEW CRYSTALS OF POTASSIUM IODIDE AND A FEW GRAINS OF SODIUM BISULFATE IN 5 ml WATER. ADD HYDROGEN PEROXIDE. SHAKE. THE FREE IODINE COLORS LIQUID BROWN.

MAKING HYDROGEN IODIDE

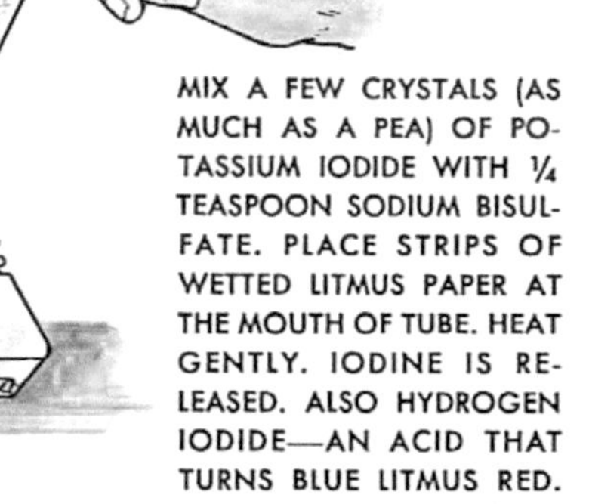

MIX A FEW CRYSTALS (AS MUCH AS A PEA) OF POTASSIUM IODIDE WITH ¼ TEASPOON SODIUM BISULFATE. PLACE STRIPS OF WETTED LITMUS PAPER AT THE MOUTH OF TUBE. HEAT GENTLY. IODINE IS RELEASED. ALSO HYDROGEN IODIDE—AN ACID THAT TURNS BLUE LITMUS RED.

STARCH TEST FOR IODINE

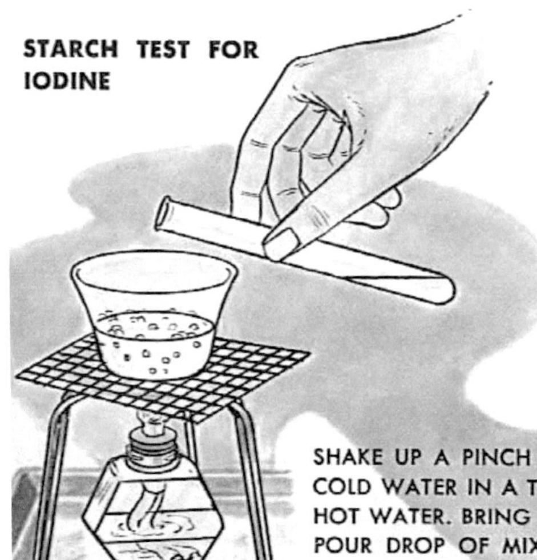

REMOVING IODINE STAIN

PAINT PAPER WITH IODINE. DISSOLVE A FEW CRYSTALS OF SODIUM THIOSULFATE ("HYPO") IN WATER. PAINT WITH THIS SOLUTION OVER THE BROWN COLOR. YOU WILL GET WHITE LETTERS AS HYPO FORMS COLORLESS COMPOUND WITH IODINE.

SHAKE UP A PINCH OF STARCH WITH COLD WATER IN A TEST TUBE. ADD TO HOT WATER. BRING TO A BOIL. COOL. POUR DROP OF MIXTURE INTO 10 ml WATER. ADD DROP OF IODINE SOLUTION. BRIGHT BLUE COLOR RESULTS.

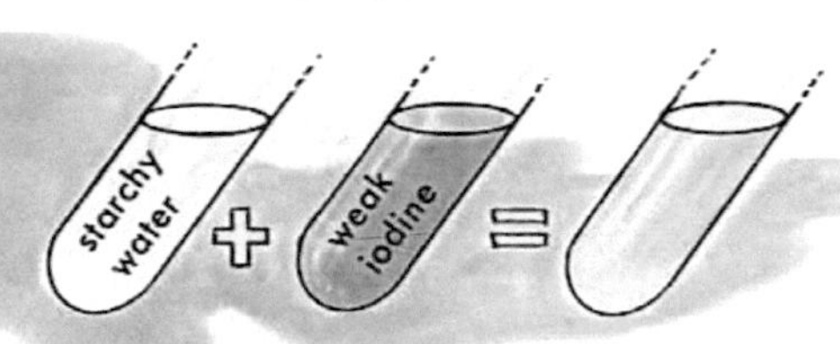

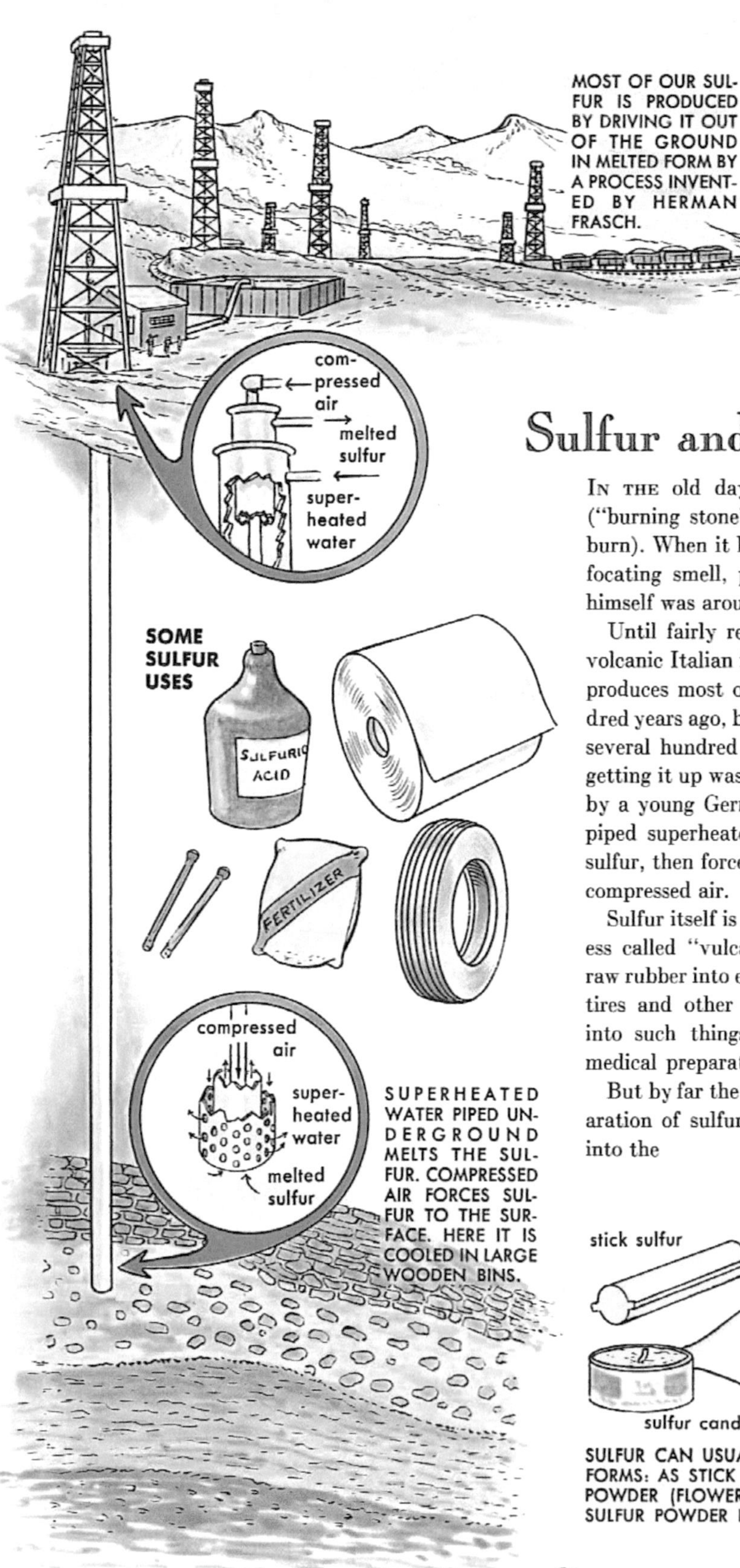

SULFUR Element 16. Atomic wt.: 32.066. Density: 2.07. Yellow crystals. Insoluble in water. Melts at 119°C. Boils at 444°C. Burns in air with blue flame.

Sulfur and Its Compounds

IN THE old days, sulfur was called "brimstone" ("burning stone" — from an old word, *brennen*, to burn). When it burned with a blue flame and a suffocating smell, people were certain that the devil himself was around.

Until fairly recently, most sulfur came from the volcanic Italian island of Sicily. But today, America produces most of the world's sulfur. About a hundred years ago, big deposits were found in Louisiana, several hundred feet underground. The problem of getting it up was solved in 1894 in a very clever way by a young German emigrant, Herman Frasch. He piped superheated water underground to melt the sulfur, then forced the melted sulfur to the top with compressed air.

Sulfur itself is used for many purposes. By a process called "vulcanization" it turns sticky, gummy raw rubber into elastic rubber usable for automobile tires and other rubber products. Sulfur also goes into such things as matches and gunpowder and medical preparations.

But by far the greatest use of sulfur is in the preparation of sulfuric acid (H_2SO_4). This acid enters into the (CONTINUED ON PAGE 52)

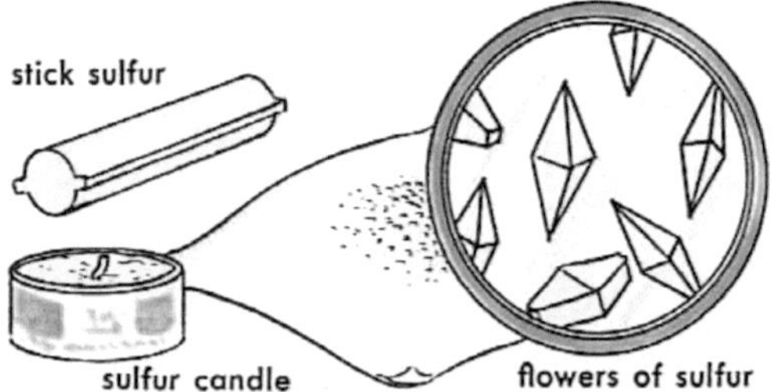

SULFUR CAN USUALLY BE BOUGHT IN THREE DIFFERENT FORMS: AS STICK SULFUR, SULFUR CANDLES, AND AS A POWDER (FLOWERS OF SULFUR). UNDER MICROSCOPE, SULFUR POWDER PROVES TO BE RHOMBIC CRYSTALS.

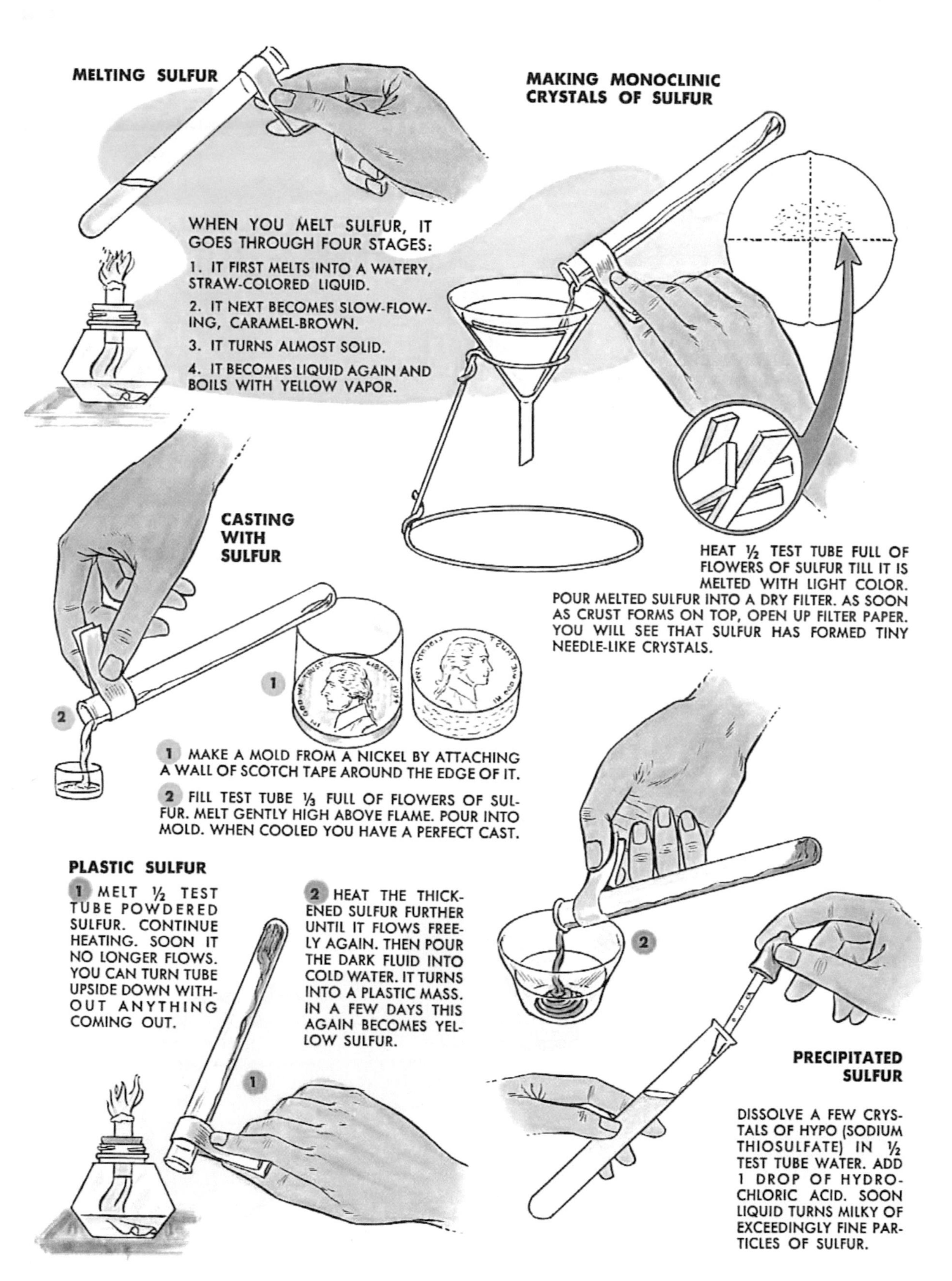
MELTING SULFUR
WHEN YOU MELT SULFUR, IT GOES THROUGH FOUR STAGES:
1. IT FIRST MELTS INTO A WATERY, STRAW-COLORED LIQUID.
2. IT NEXT BECOMES SLOW-FLOWING, CARAMEL-BROWN.
3. IT TURNS ALMOST SOLID.
4. IT BECOMES LIQUID AGAIN AND BOILS WITH YELLOW VAPOR.
MAKING MONOCLINIC CRYSTALS OF SULFUR
HEAT ½ TEST TUBE FULL OF FLOWERS OF SULFUR TILL IT IS MELTED WITH LIGHT COLOR. POUR MELTED SULFUR INTO A DRY FILTER. AS SOON AS CRUST FORMS ON TOP, OPEN UP FILTER PAPER. YOU WILL SEE THAT SULFUR HAS FORMED TINY NEEDLE-LIKE CRYSTALS.
CASTING WITH SULFUR
1
2
1 MAKE A MOLD FROM A NICKEL BY ATTACHING A WALL OF SCOTCH TAPE AROUND THE EDGE OF IT.
2 FILL TEST TUBE ⅓ FULL OF FLOWERS OF SULFUR. MELT GENTLY HIGH ABOVE FLAME. POUR INTO MOLD. WHEN COOLED YOU HAVE A PERFECT CAST.
PLASTIC SULFUR
1 MELT ½ TEST TUBE POWDERED SULFUR. CONTINUE HEATING. SOON IT NO LONGER FLOWS. YOU CAN TURN TUBE UPSIDE DOWN WITHOUT ANYTHING COMING OUT.
2 HEAT THE THICKENED SULFUR FURTHER UNTIL IT FLOWS FREELY AGAIN. THEN POUR THE DARK FLUID INTO COLD WATER. IT TURNS INTO A PLASTIC MASS. IN A FEW DAYS THIS AGAIN BECOMES YELLOW SULFUR.
1
2
PRECIPITATED SULFUR
DISSOLVE A FEW CRYSTALS OF HYPO (SODIUM THIOSULFATE) IN ½ TEST TUBE WATER. ADD 1 DROP OF HYDROCHLORIC ACID. SOON LIQUID TURNS MILKY OF EXCEEDINGLY FINE PARTICLES OF SULFUR.

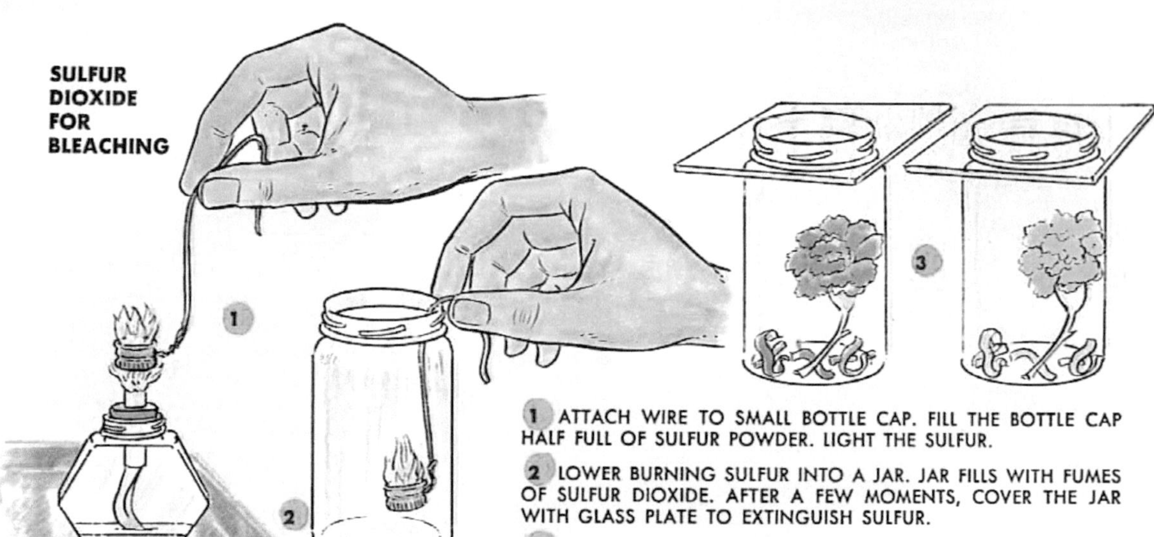

1 ATTACH WIRE TO SMALL BOTTLE CAP. FILL THE BOTTLE CAP HALF FULL OF SULFUR POWDER. LIGHT THE SULFUR.

2 LOWER BURNING SULFUR INTO A JAR. JAR FILLS WITH FUMES OF SULFUR DIOXIDE. AFTER A FEW MOMENTS, COVER THE JAR WITH GLASS PLATE TO EXTINGUISH SULFUR.

3 LIFT GLASS PLATE. DROP INTO JAR APPLE PEELINGS AND MOISTENED, BRIGHT-COLORED FLOWER. COVER AGAIN WITH GLASS PLATE. IN A SHORT WHILE, COLORS HAVE BLEACHED.

Be careful not to breathe fumes.

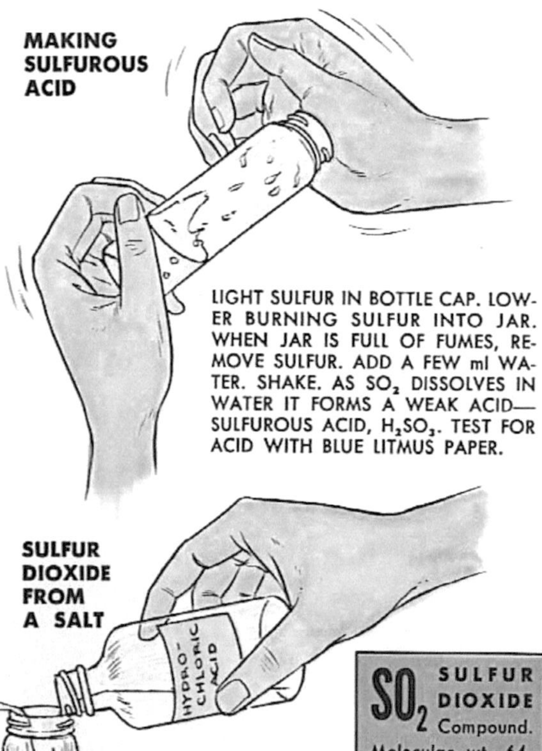

LIGHT SULFUR IN BOTTLE CAP. LOWER BURNING SULFUR INTO JAR. WHEN JAR IS FULL OF FUMES, REMOVE SULFUR. ADD A FEW ml WATER. SHAKE. AS SO_2 DISSOLVES IN WATER IT FORMS A WEAK ACID—SULFUROUS ACID, H_2SO_3. TEST FOR ACID WITH BLUE LITMUS PAPER.

DISSOLVE ½ TEASPOON HYPO (SODIUM THIOSULFATE) IN 40 ml WATER. ADD A FEW ml HYDROCHLORIC ACID. SULFUR DIOXIDE AND PRECIPITATE OF SULFUR RESULT.

SO_2 SULFUR DIOXIDE Compound. Molecular wt. 64. Colorless gas with a choking odor. Does not burn nor support combustion. 2.2 weight of air. Highly soluble in water—3,937 vols. in 100 vols. at 20° C.

Sulfur—Continued

production — directly or indirectly — of practically every manufactured article we use today. It is used in refining gasoline, in making steel and paper, fibers and films, plastics and explosives, and thousands of other chemicals.

Sulfur Dioxide— The first step in making sulfuric acid from sulfur is to burn the sulfur.

When burning in the air, each atom of sulfur takes on two atoms of oxygen to make one molecule of sulfur dioxide gas (SO_2).

By a special, complicated process, sulfur dioxide can be forced to take on another oxygen atom and form sulfur trioxide (SO_3). With water, this makes sulfuric acid:

$$H_2O + SO_3 \rightarrow H_2SO_4$$

Hydrogen Sulfide— Many sulfur compounds have unpleasant, penetrating smells. Some of these compounds have very complex molecules — just imagine a skunk producing a chemical with this formula: $CH_3CH_2CH_2CH_2SH$! The smell of rotten eggs, on the other hand, comes from the simple compound hydrogen sulfide (H_2S).

Hydrogen sulfide is used in chemical analysis to determine what metals are found in a certain substance. It combines with metals into salts (sulfides) that can be distinguished from each other by their colors and by the way they react with acids and other chemicals.

NOTE: Perform these experiments out-of-doors or before an open window. Be careful not to breathe fumes.

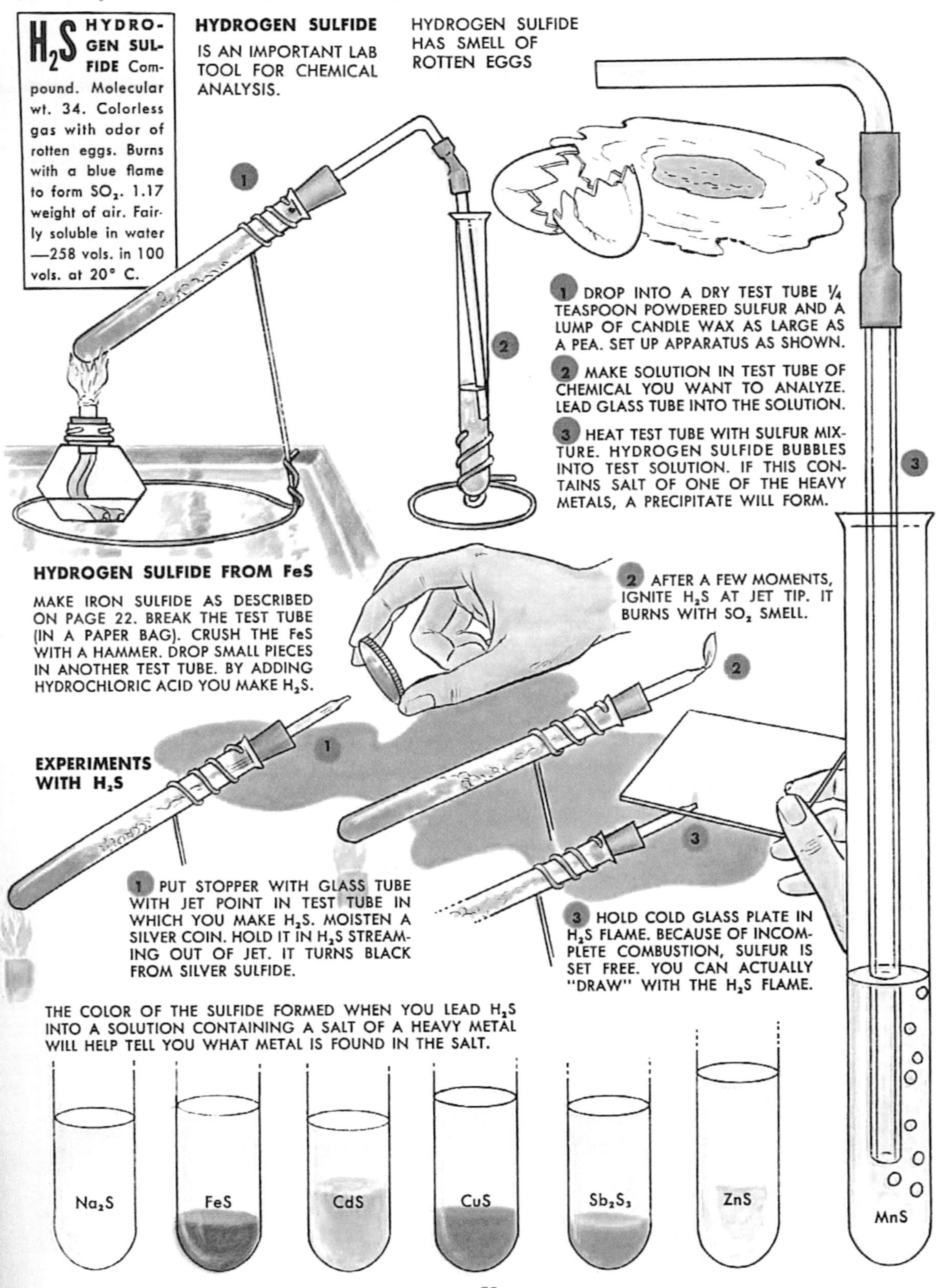

H_2S HYDROGEN SULFIDE Compound. Molecular wt. 34. Colorless gas with odor of rotten eggs. Burns with a blue flame to form SO_2. 1.17 weight of air. Fairly soluble in water —258 vols. in 100 vols. at 20° C.

HYDROGEN SULFIDE IS AN IMPORTANT LAB TOOL FOR CHEMICAL ANALYSIS.

HYDROGEN SULFIDE HAS SMELL OF ROTTEN EGGS

1 DROP INTO A DRY TEST TUBE ¼ TEASPOON POWDERED SULFUR AND A LUMP OF CANDLE WAX AS LARGE AS A PEA. SET UP APPARATUS AS SHOWN.

2 MAKE SOLUTION IN TEST TUBE OF CHEMICAL YOU WANT TO ANALYZE. LEAD GLASS TUBE INTO THE SOLUTION.

3 HEAT TEST TUBE WITH SULFUR MIXTURE. HYDROGEN SULFIDE BUBBLES INTO TEST SOLUTION. IF THIS CONTAINS SALT OF ONE OF THE HEAVY METALS, A PRECIPITATE WILL FORM.

HYDROGEN SULFIDE FROM FeS

MAKE IRON SULFIDE AS DESCRIBED ON PAGE 22. BREAK THE TEST TUBE (IN A PAPER BAG). CRUSH THE FeS WITH A HAMMER. DROP SMALL PIECES IN ANOTHER TEST TUBE. BY ADDING HYDROCHLORIC ACID YOU MAKE H_2S.

EXPERIMENTS WITH H_2S

1 PUT STOPPER WITH GLASS TUBE WITH JET POINT IN TEST TUBE IN WHICH YOU MAKE H_2S. MOISTEN A SILVER COIN. HOLD IT IN H_2S STREAMING OUT OF JET. IT TURNS BLACK FROM SILVER SULFIDE.

2 AFTER A FEW MOMENTS, IGNITE H_2S AT JET TIP. IT BURNS WITH SO_2 SMELL.

3 HOLD COLD GLASS PLATE IN H_2S FLAME. BECAUSE OF INCOMPLETE COMBUSTION, SULFUR IS SET FREE. YOU CAN ACTUALLY "DRAW" WITH THE H_2S FLAME.

THE COLOR OF THE SULFIDE FORMED WHEN YOU LEAD H_2S INTO A SOLUTION CONTAINING A SALT OF A HEAVY METAL WILL HELP TELL YOU WHAT METAL IS FOUND IN THE SALT.

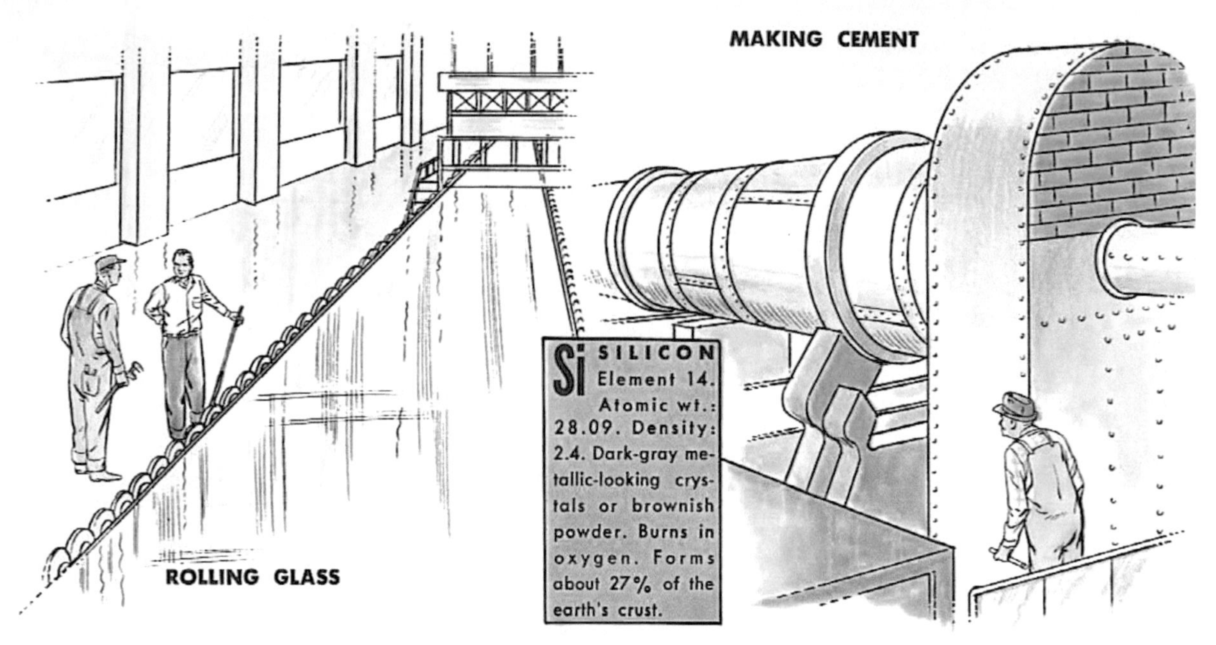

Silicon—The Element You Step On

SILICON (from the Latin *silex*, flint) is the second most abundant element on earth — after oxygen. Whether you are walking on sand or clay, rock or cement, almost half of what you're stepping on is silicon.

Silicon is found in nature in combination with oxygen (mostly the dioxide, SiO_2) and in different silicates (salts of various silicic acids).

With few exceptions, silicon compounds are insoluble in water. And that is a good thing for all of us. The glass of our windows and the glasses from which we drink are silicates. So are the glazes on our cups and the enamel on our bathtubs. Most glass and many glazes are made by fusing together sand (SiO_2), limestone, and soda.

The silicates of sodium and potassium dissolve in water. A concentrated solution of sodium silicate (Na_2SiO_3) is sold in hardware stores under the name of waterglass. It is used as a glue, for fireproofing wood and for preserving eggs.

Within recent years, chemists have developed a whole line of new silicon compounds called silicones. Some of them are oil-like. Some look like putty ("Silly Putty"). Still others are rubber-like. Paper and cloth can be made water-repellent by being treated with suitable silicones.

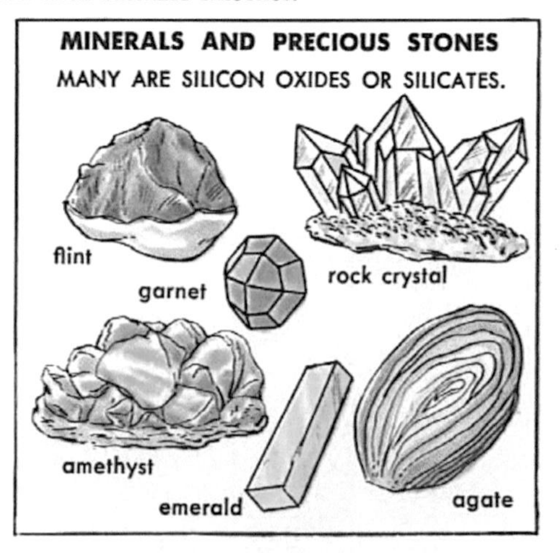

MAKING SILICIC ACID

1 IN ONE GLASS, DILUTE 20 ml WATERGLASS WITH 20 ml WATER.

2 IN ANOTHER, MIX 10 ml HYDROCHLORIC ACID AND 10 ml WATER.

3 POUR THE TWO MIXTURES AT ONE TIME INTO A THIRD GLASS.

4 STAND SPOON UPRIGHT IN THE MIXTURE WHICH, ALMOST IMMEDIATELY, TURNS INTO A JELLY ("GEL") SO STIFF THAT SPOON STANDS BY ITSELF AND YOU CAN TURN THE GLASS UPSIDE DOWN.

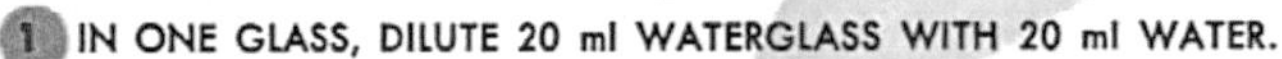

MAKING SILICON DIOXIDE

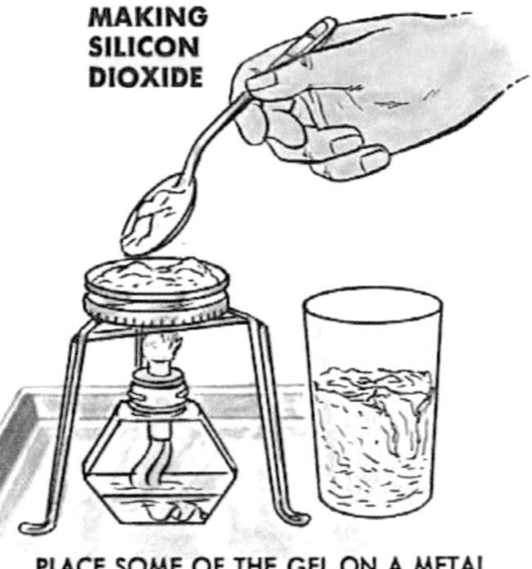

PLACE SOME OF THE GEL ON A METAL JAR LID. HEAT. THE SILICIC ACID (H_2SiO_3) GIVES UP WATER (H_2O) AND TURNS INTO A GRAYISH-WHITE POWDER OF SILICON DIOXIDE (SiO_2).

WEAKNESS OF SILICIC ACID

MAKING WATER-GLASS

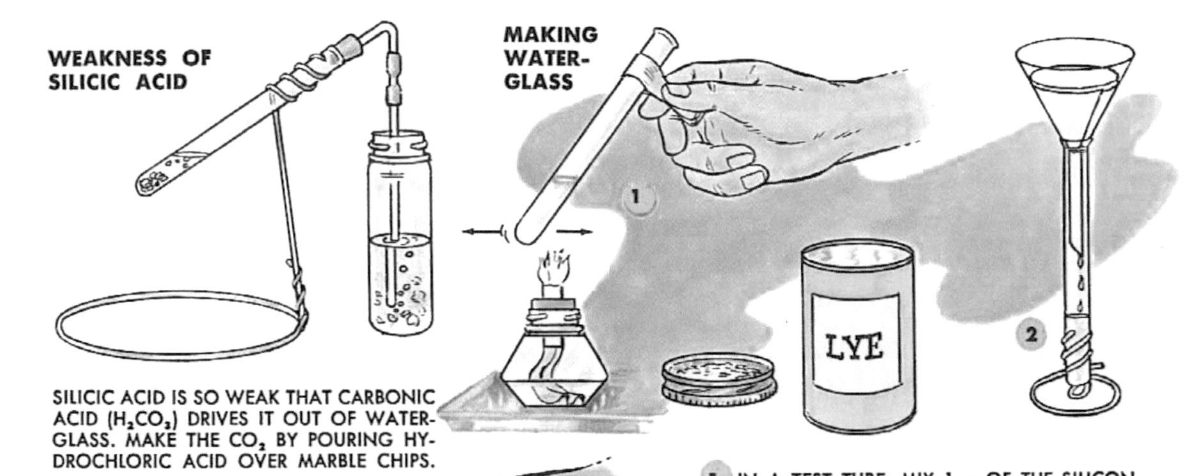

SILICIC ACID IS SO WEAK THAT CARBONIC ACID (H_2CO_3) DRIVES IT OUT OF WATERGLASS. MAKE THE CO_2 BY POURING HYDROCHLORIC ACID OVER MARBLE CHIPS.

1 IN A TEST TUBE, MIX 1 g OF THE SILICON DIOXIDE YOU MADE, 2 g LYE (NaOH), AND 5 ml WATER. HEAT CAREFULLY, MOVING TUBE.

2 AFTER FILTERING, YOU WILL HAVE A CLEAR SOLUTION OF SODIUM SILICATE (Na_2SiO_3).

MAKING SILICATES

1 DILUTE 5 ml WATERGLASS (Na_2SiO_3) WITH 5 ml WATER.

2 DISSOLVE SMALL CRYSTAL OF COPPER SULFATE IN WATER.

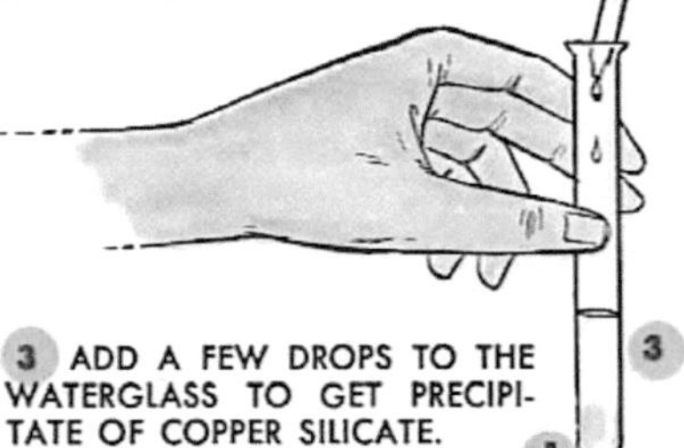

3 ADD A FEW DROPS TO THE WATERGLASS TO GET PRECIPITATE OF COPPER SILICATE.

"GROWING" A SILICON "JUNGLE"

IN A PINT JAR, PLACE ½-INCH LAYER OF SAND. POUR ON TOP OF THIS A MIXTURE OF EQUAL PARTS OF WATERGLASS AND WATER. PLACE IT IN A SPOT WHERE IT WILL NOT BE DISTURBED. DROP IN CRYSTALS OF VARIOUS SALTS YOU MAY HAVE: IRON SULFATE, COPPER SULFATE, ALUM, EPSOM SALT. THE CRYSTALS SEND UP "SHOOTS." IN A FEW HOURS, YOUR SILICATE "JUNGLE" IS FULLY "GROWN."

Boron—Future Rocket-Power Element?

Less than a hundred years ago, a mineral called borax, containing the element boron, was carted out of Death Valley in California by twenty-mule teams — about the slowest transportation you can think of. Someday, boron may be put in zip-fuels for space missiles — the fastest form of transportation imaginable. Boron has the ability (as does carbon) to combine with hydrogen in a number of ways. When these boranes or boron hydrides burn, they develop a tremendous amount of power.

Boron can be isolated as a hard, brownish-black powder. Its carbon compound, boron carbide (B_4C), is almost as hard as diamond.

But you are probably more familiar with boron

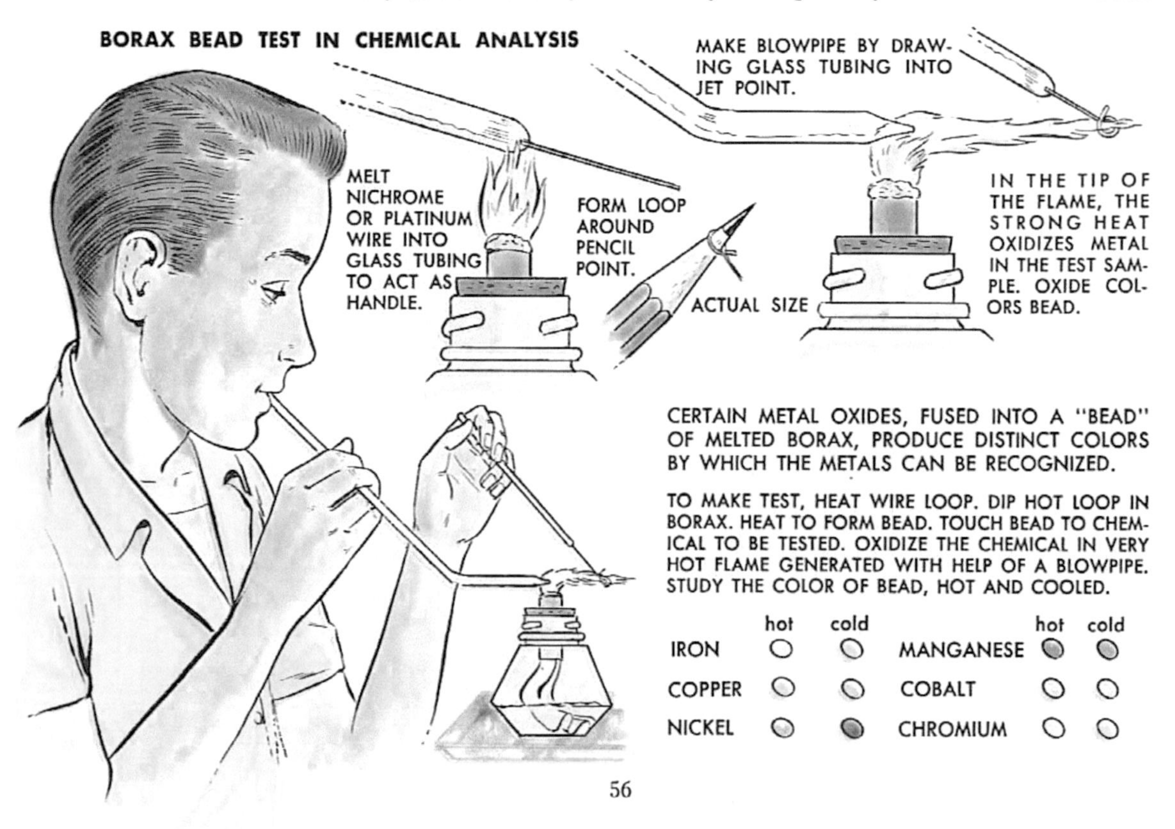

through two of its compounds which are found in almost every household: boric acid (H_3BO_3), used as a mild antiseptic, and borax (sodium tetraborate, $Na_2B_4O_7 \cdot 10H_2O$), used for cleaning purposes and as a water softener.

Borax has a great number of uses outside the home. It is used for soldering, for producing certain kinds of soap, and for making other boron compounds.

The glass industry uses large quantities of borax for making boron-aluminum-silicate glass. You know this kind of glass by its trade name, Pyrex. Kitchen utensils and laboratory ware made of Pyrex glass have the great advantage over ordinary glass that they can be placed directly on the fire and do not break so easily when they are subjected to sudden heating or cooling.

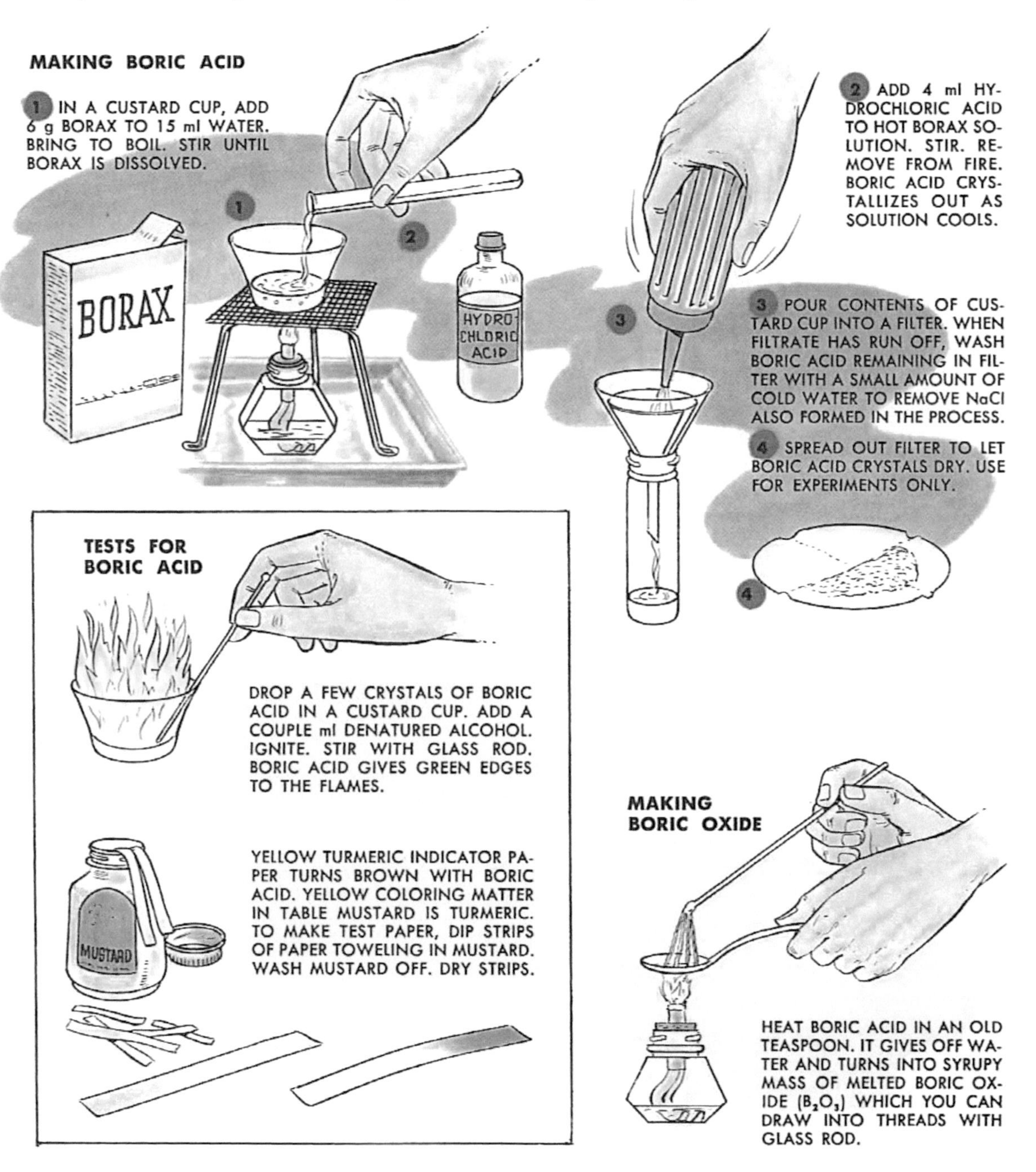

Sodium and Potassium

THE SALTS of sodium and potassium have been used for thousands of years in making soap and glass and for a great number of other purposes.

Sodium chloride (NaCl) is the most common sodium salt — it is the chemical that makes ocean water "salty." Plants growing in the ocean take up so much of the sodium that people along the seacoasts of the world used to burn dried seaweed to secure "soda ash" (sodium carbonate, Na_2CO_3). Inland plants, on the other hand, pick up potassium from the soil. Inland people boiled out wood ashes in large pots to get "potash" (potassium carbonate, K_2CO_3).

In 1807, the British scientist, Humphry Davy, succeeded in isolating the metals found in these salts. They proved to be wax-soft and silvery. He called them sodium (from soda ash) and potassium (from potash). These are still their English names. But in chemical formulas they are referred to as natrium (Na) and kalium (K) — from abbreviations of the Arabic names of the ashes: *natrun* and *al qili* (alkali).

Na NATRIUM Element 11. Atomic wt.: 22.991. Density: 0.97. (English: Sodium) Silver-white metal, can be cut with knife. Oxidizes in air. Reacts with water. Burns with yellow flame.

"SALARY" COMES FROM *SALARIUM* —THE WAGES PAID IN SALT TO ROMAN SOLDIERS.

CRYSTALLIZING SALT BY EVAPORATION

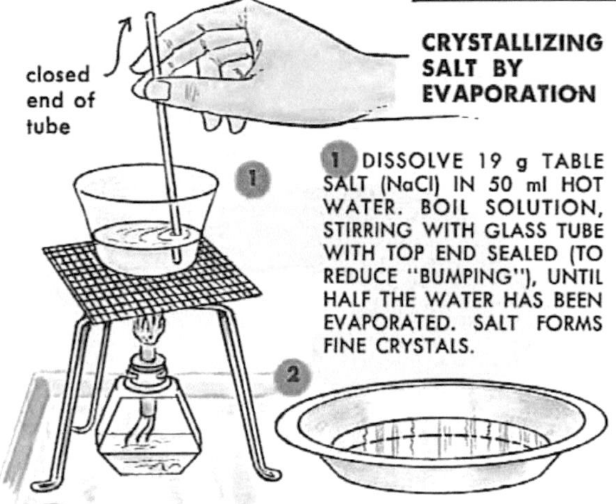

1 DISSOLVE 19 g TABLE SALT (NaCl) IN 50 ml HOT WATER. BOIL SOLUTION, STIRRING WITH GLASS TUBE WITH TOP END SEALED (TO REDUCE "BUMPING"), UNTIL HALF THE WATER HAS BEEN EVAPORATED. SALT FORMS FINE CRYSTALS.

2 POUR CLEAR LIQUID INTO LARGE PIE PLATE. PLACE IN SUNNY WINDOW FOR WATER TO EVAPORATE SLOWLY. THE CRYSTALS FORMED WILL BE MUCH LARGER.

MAKING NORMAL SALT FROM ACID SALT

SODIUM SULFATE IS PRODUCED BY HEATING SODIUM ACID SULFATE WITH SODIUM CHLORIDE.

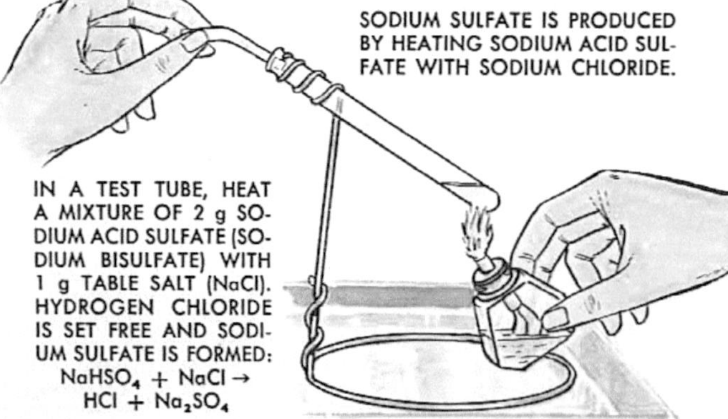

IN A TEST TUBE, HEAT A MIXTURE OF 2 g SODIUM ACID SULFATE (SODIUM BISULFATE) WITH 1 g TABLE SALT (NaCl). HYDROGEN CHLORIDE IS SET FREE AND SODIUM SULFATE IS FORMED:

$$NaHSO_4 + NaCl \rightarrow HCl + Na_2SO_4$$

MAKING ACID SALT FROM NORMAL SALT

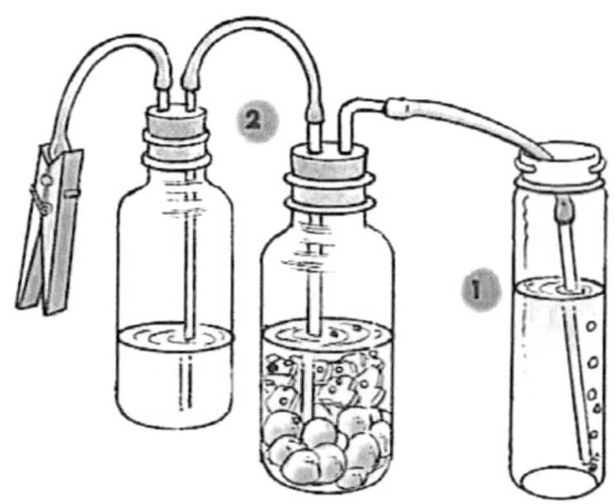

THE ACID CARBONATE ($NaHCO_3$) IS MADE BY LEADING CO_2 TO NORMAL CARBONATE (Na_2CO_3).

1 MAKE SATURATED SOLUTION BY SHAKING 3 TEASPOONS WASHING SODA IN 30 ml COOL, BUT NOT COLD, WATER. FILTER IT.

2 SET UP APPARATUS FOR MAKING CO_2 AS SHOWN ON PAGE 31. LEAD CO_2 INTO SODA SOLUTION FOR 10 MINUTES. THEN SET ASIDE. SHORTLY $NaHCO_3$ CRYSTALS APPEAR.

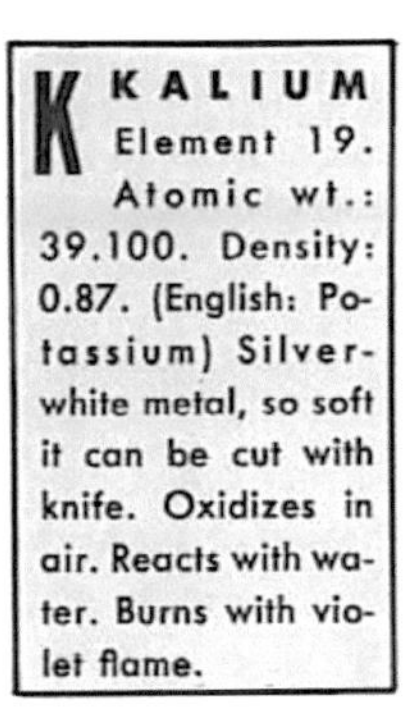

K KALIUM Element 19. Atomic wt.: 39.100. Density: 0.87. (English: Potassium) Silver-white metal, so soft it can be cut with knife. Oxidizes in air. Reacts with water. Burns with violet flame.

PIONEER WOMEN MADE POTASH FROM WOOD ASHES.

NITRATE TO NITRITE

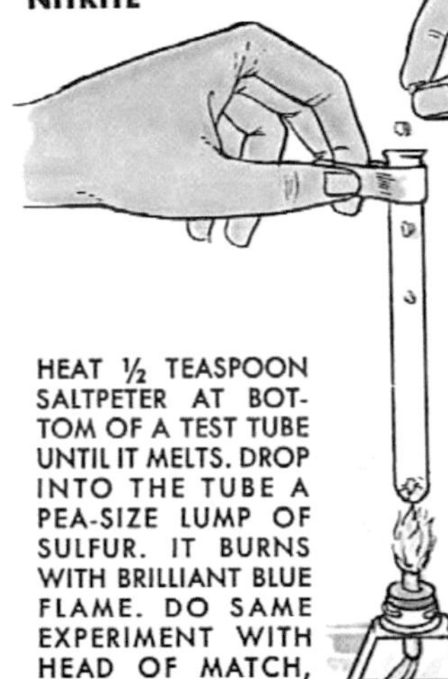

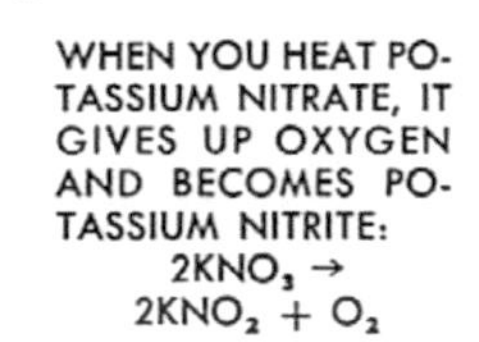

WHEN YOU HEAT POTASSIUM NITRATE, IT GIVES UP OXYGEN AND BECOMES POTASSIUM NITRITE:

$$2KNO_3 \rightarrow 2KNO_2 + O_2$$

HEAT ½ TEASPOON SALTPETER AT BOTTOM OF A TEST TUBE UNTIL IT MELTS. DROP INTO THE TUBE A PEA-SIZE LUMP OF SULFUR. IT BURNS WITH BRILLIANT BLUE FLAME. DO SAME EXPERIMENT WITH HEAD OF MATCH, CHARCOAL BIT.

MAKING POTASH

1 STIR UP SEVERAL TEASPOONS OF FRESH WOOD ASHES WITH WARM WATER. SKIM OFF WOOD REMAINS.

2 FILTER THE MIXTURE OF ASHES AND WATER. COLLECT FILTRATE IN A CUSTARD CUP. EVAPORATE MOST OF WATER. THEN COOL TO PERMIT K_2CO_3 TO CRYSTALLIZE OUT.

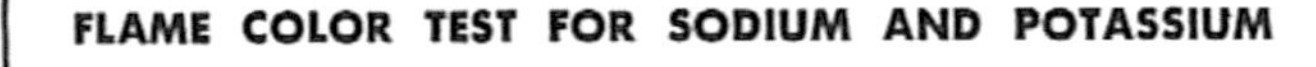

FLAME COLOR TEST FOR SODIUM AND POTASSIUM

THE COMPOUNDS OF CERTAIN METALS GIVE DISTINCT COLORS TO A FLAME. DIP NICHROME WIRE IN HCl TO CLEAN IT. HEAT IT. THEN DIP LOOP IN COMPOUND AND HOLD IN FLAME.

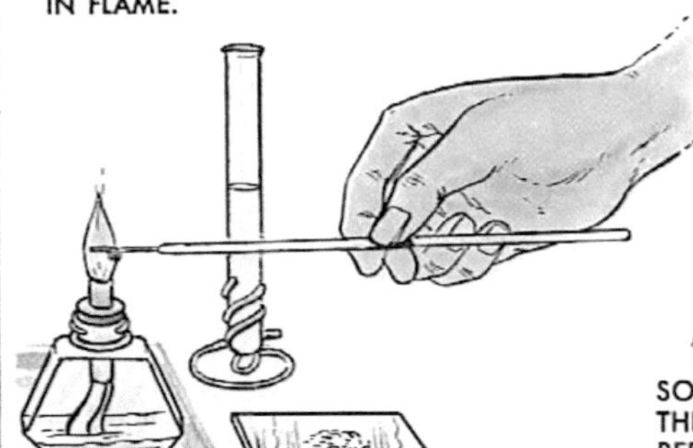

SODIUM COMPOUNDS GIVE THE FLAME A BRIGHT YELLOW-RED COLOR. POTASSIUM COMPOUNDS GIVE VIOLET FLAME.

TO SEE VIOLET COLOR OF POTASSIUM IN MIXTURE WITH Na, USE BLUE GLASS TO SCREEN OUT YELLOW OF Na.

Calcium— for Building

STAND UP STRAIGHT. You can do it because your bones contain calcium. Tell a mason to put up a brick house. He can do it with mortar containing calcium. Tell a master builder to build a monument. He will make it from marble — calcium again. Tell a hen to "go lay an egg." She can do it if she gets enough calcium in her feed to make the shell.

Calcium carbonate ($CaCO_3$) is the starting point for most calcium compounds — and for other chemicals as well. It is found in nature in cliffs and mountain ranges in the form of chalk and limestone and marble. And it makes up the shells of clams and mussels and billions of tiny sea creatures.

Calcium carbonate is almost insoluble in water. But if the water contains carbon dioxide, some goes in solution as calcium bicarbonate ($Ca(HCO_3)_2$). This explains the formations in our famous limestone caves. Rainwater containing carbon dioxide seeped through the ground and dissolved a small amount of limestone. In falling from the cave ceiling and drying, the drops gave up H_2O and CO_2 and left $CaCO_3$ behind. The minute deposits of falling drops during thousands of years created the stalactites hanging from the roof of the caves and the stalagmites rising from the floor.

A widespread mineral called gypsum is the sulfate of calcium. In this, each molecule of sulfate has two molecules of water attached to it ($CaSO_4 \cdot 2H_2O$). When gypsum is heated, it loses three quarters of its water and becomes plaster of Paris ($2CaSO_4 \cdot H_2O$). When you mix plaster of Paris and water, it again takes on the full amount of H_2O and hardens into a hydrate similar to the original gypsum.

The name of calcium was given to the metal hidden in limestone by its discoverer, Humphry Davy. It comes from *calx*, the old Latin name for lime.

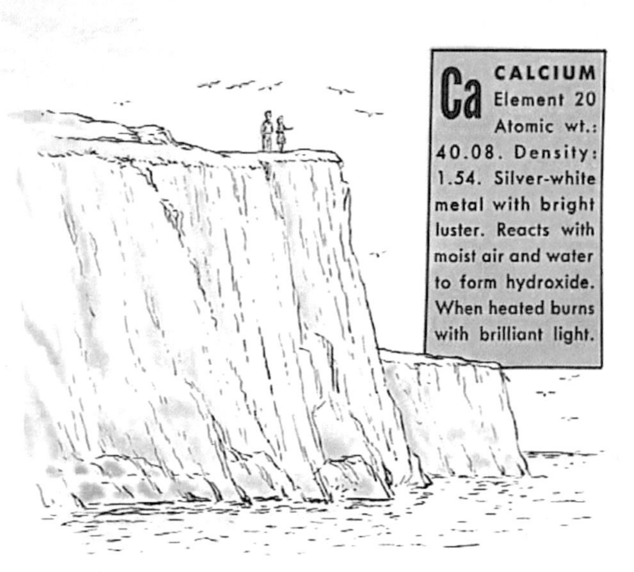

"THE WHITE CLIFFS OF DOVER" CONSIST OF ALMOST PURE CALCIUM CARBONATE IN THE FORM OF CHALK.

STALACTITES AND STALAGMITES ARE UNDERGROUND DEPOSITS OF $CaCO_3$.

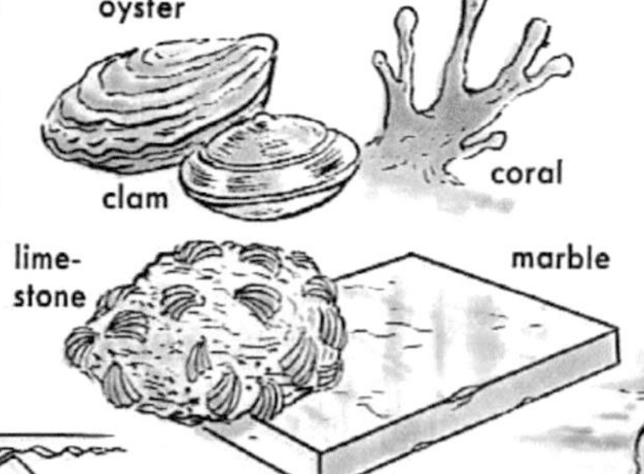

SEA SHELLS, CORAL, LIMESTONE, AND MARBLE ARE ALL CALCIUM CARBONATE.

WHEN LIMESTONE IS HEATED IN KILNS, IT LOSES CARBON DIOXIDE AND TURNS INTO QUICKLIME—CALCIUM OXIDE.

WHEN WATER IS ADDED TO LUMPS OF QUICKLIME (CaO), THEY CRUMBLE INTO A WHITISH POWDER OF SLAKED OR HYDRATED LIME ($Ca(OH)_2$). (SEE ALSO PAGE 45.)

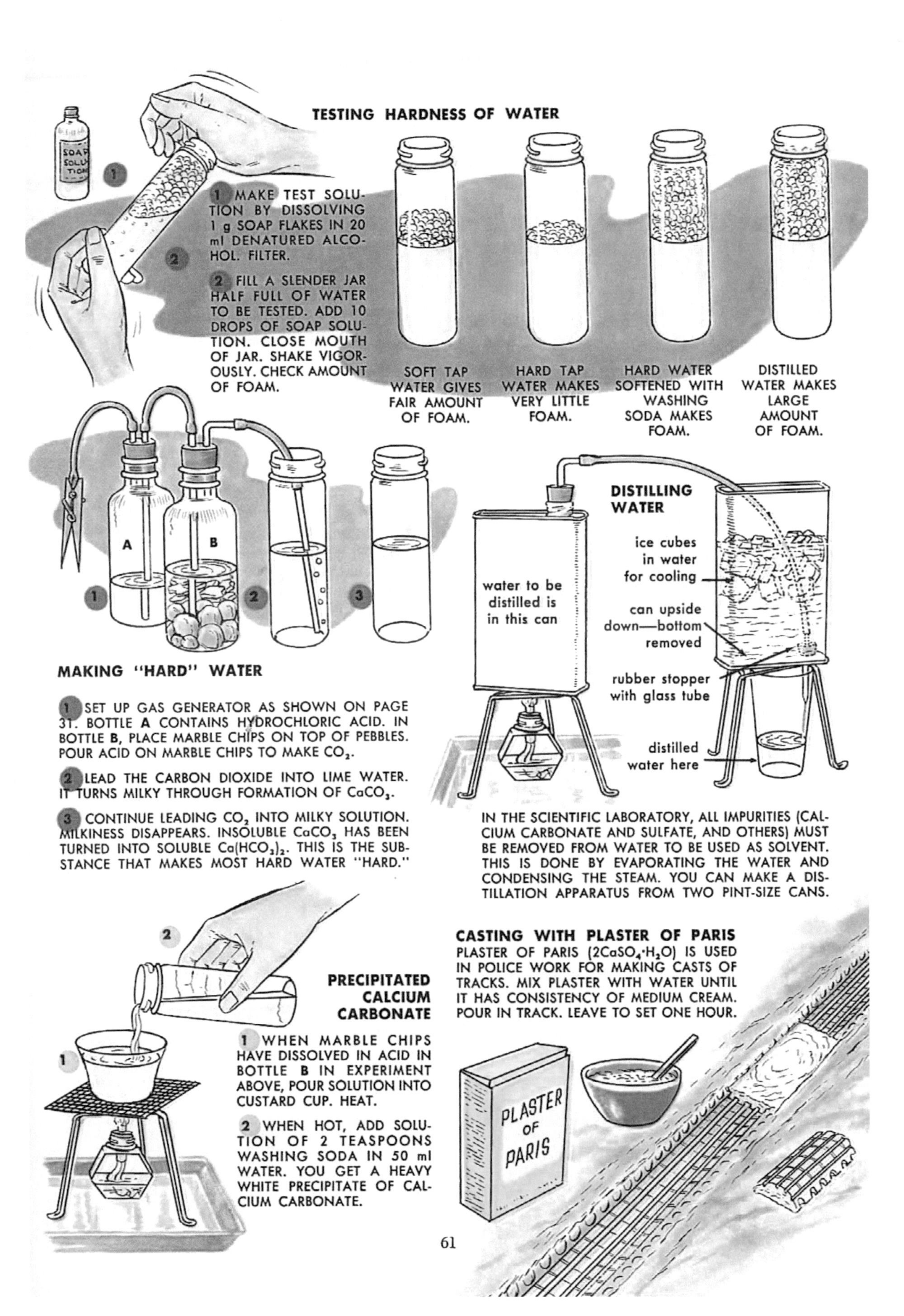

TESTING HARDNESS OF WATER

1 MAKE TEST SOLUTION BY DISSOLVING 1 g SOAP FLAKES IN 20 ml DENATURED ALCOHOL. FILTER.

2 FILL A SLENDER JAR HALF FULL OF WATER TO BE TESTED. ADD 10 DROPS OF SOAP SOLUTION. CLOSE MOUTH OF JAR. SHAKE VIGOROUSLY. CHECK AMOUNT OF FOAM.

SOFT TAP WATER GIVES FAIR AMOUNT OF FOAM.

HARD TAP WATER MAKES VERY LITTLE FOAM.

HARD WATER SOFTENED WITH WASHING SODA MAKES FOAM.

DISTILLED WATER MAKES LARGE AMOUNT OF FOAM.

MAKING "HARD" WATER

1 SET UP GAS GENERATOR AS SHOWN ON PAGE 31. BOTTLE **A** CONTAINS HYDROCHLORIC ACID. IN BOTTLE **B**, PLACE MARBLE CHIPS ON TOP OF PEBBLES. POUR ACID ON MARBLE CHIPS TO MAKE CO_2.

2 LEAD THE CARBON DIOXIDE INTO LIME WATER. IT TURNS MILKY THROUGH FORMATION OF $CaCO_3$.

3 CONTINUE LEADING CO_2 INTO MILKY SOLUTION. MILKINESS DISAPPEARS. INSOLUBLE $CaCO_3$ HAS BEEN TURNED INTO SOLUBLE $Ca(HCO_3)_2$. THIS IS THE SUBSTANCE THAT MAKES MOST HARD WATER "HARD."

DISTILLING WATER

IN THE SCIENTIFIC LABORATORY, ALL IMPURITIES (CALCIUM CARBONATE AND SULFATE, AND OTHERS) MUST BE REMOVED FROM WATER TO BE USED AS SOLVENT. THIS IS DONE BY EVAPORATING THE WATER AND CONDENSING THE STEAM. YOU CAN MAKE A DISTILLATION APPARATUS FROM TWO PINT-SIZE CANS.

PRECIPITATED CALCIUM CARBONATE

1 WHEN MARBLE CHIPS HAVE DISSOLVED IN ACID IN BOTTLE **B** IN EXPERIMENT ABOVE, POUR SOLUTION INTO CUSTARD CUP. HEAT.

2 WHEN HOT, ADD SOLUTION OF 2 TEASPOONS WASHING SODA IN 50 ml WATER. YOU GET A HEAVY WHITE PRECIPITATE OF CALCIUM CARBONATE.

CASTING WITH PLASTER OF PARIS

PLASTER OF PARIS ($2CaSO_4 \cdot H_2O$) IS USED IN POLICE WORK FOR MAKING CASTS OF TRACKS. MIX PLASTER WITH WATER UNTIL IT HAS CONSISTENCY OF MEDIUM CREAM. POUR IN TRACK. LEAVE TO SET ONE HOUR.

Let's Compare Two Metals

HOLD 2-INCH PIECE OF MAGNESIUM RIBBON WITH A PAIR OF PLIERS. IGNITE IT. IT BURNS WITH A BRILLIANT, WHITE FLAME. MIX ASHES (MgO) WITH WATER. TEST MIXTURE WITH RED LITMUS PAPER.

CUT SLIVER OF ZINC. HOLD IT IN FLAME. IT BURNS WITH BLUISH-GREEN FLAME TO ZINC OXIDE. ZnO IS YELLOW WHEN HOT, WHITE WHEN COOL.

TAKE A LOOK at the periodic table of elements on pages 38-39. In column IIA you find the metal magnesium, in column IIB the metal zinc. The fact that the two families in which they are found both have the Roman numeral II would indicate that they are related. But the fact that they are in separate "subgroups" would suggest that they differ in certain respects. That is exactly the case.

In their compounds they are very much alike. One atom combines with one atom of oxygen to form the oxide (MgO and ZnO), and one atom replaces two atoms of hydrogen in forming a salt ($MgCl_2$ and $ZnCl_2$, for instance). But in some of their reactions they do not behave alike — as you will learn.

Before World War II, magnesium had little use — mainly in flash photography because it burns with a blinding, white light. But the metal became important when lightweight planes were needed — melted together with other metals it forms an "alloy"

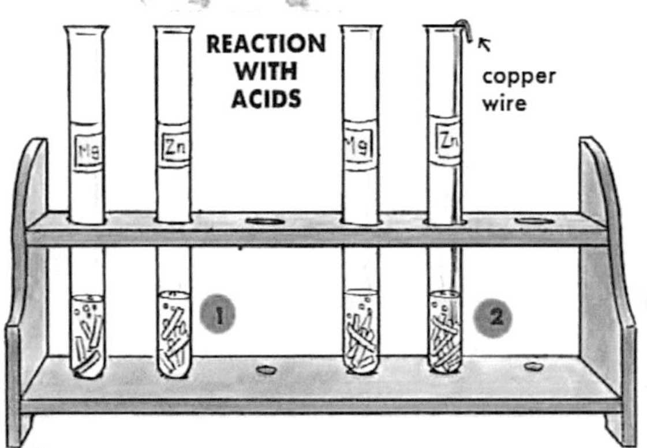

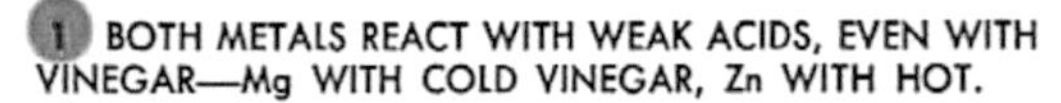

1 BOTH METALS REACT WITH WEAK ACIDS, EVEN WITH VINEGAR—Mg WITH COLD VINEGAR, Zn WITH HOT.

2 POUR SOLUTION OF 1 g SODIUM BISULFATE IN 10 ml WATER ON Mg AND Zn. Mg REACTS FAST, Zn SLOWLY. NOW TOUCH ZINC WITH A COPPER WIRE. REACTION SPEEDS UP, CAUSED BY ELECTRIC PROCESS.

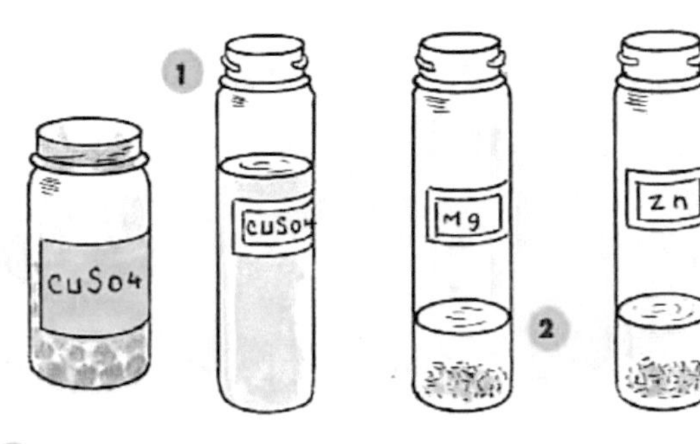

1 DISSOLVE 4 g COPPER SULFATE IN 40 ml WATER. POUR HALF OF THE SOLUTION OVER STRIPS OF MAGNESIUM, THE OTHER HALF OVER SLIVERS OF ZINC.

2 COPPER IS FORCED OUT AND Mg AND Zn GO INTO SOLUTION. IF ENOUGH METAL IS USED, THE BLUE COLOR DISAPPEARS. $MgSO_4$ AND $ZnSO_4$ ARE COLORLESS.

Mg MAGNESIUM Element 12. Atomic wt.: 24.32. Density: 1.75. Silver-white metal. Ductile, malleable. Reacts with boiling water. Burns in air with very brilliant white light.	**Zn** ZINC Element 30. Atomic wt.: 65.38. Density: 7.1. Bluish-white metal. Ductile and malleable. Distils when heated to boiling. Can be made to burn with bluish flame.

that is light yet very strong. Some magnesium compounds are used in medicine: milk of magnesia ($Mg(OH)_2$) and Epsom salt ($MgSO_4 \cdot 7H_2O$).

Zinc has been used for ages to coat iron pails and pipes to prevent them from rusting — "galvanized iron." Zinc is also a part of many alloys (German silver and brass) and is important in the making of dry-cell batteries.

MAKING THE HYDROXIDES

1 ADD SODIUM HYDROXIDE SOLUTION TO SOLUTION OF MAGNESIUM SULFATE. WHITE $Mg(OH)_2$ FORMS.

2 ADD SMALL AMOUNT OF NaOH SOLUTION TO DILUTED TINNERS' FLUID ($ZnCl_2$). $Zn(OH)_2$ IS FORMED. ADD MORE NaOH. PRECIPITATE DISSOLVES WITH FORMATION OF SODIUM ZINCATE ($Na_2Zn\ O_2$).

3 ADD AMMONIA (AMMONIUM HYDROXIDE) TO MAGNESIUM SULFATE SOLUTION. AGAIN $Mg(OH)_2$ FORMS.

4 ADD SMALL AMOUNT OF AMMONIA TO DILUTED TINNERS' FLUID. $Zn(OH)_2$ FORMS. ADD MORE. THE $Zn(OH)_2$ DISSOLVES, FORMING COMPOUND WITH NH_3.

MAKING THE CARBONATES

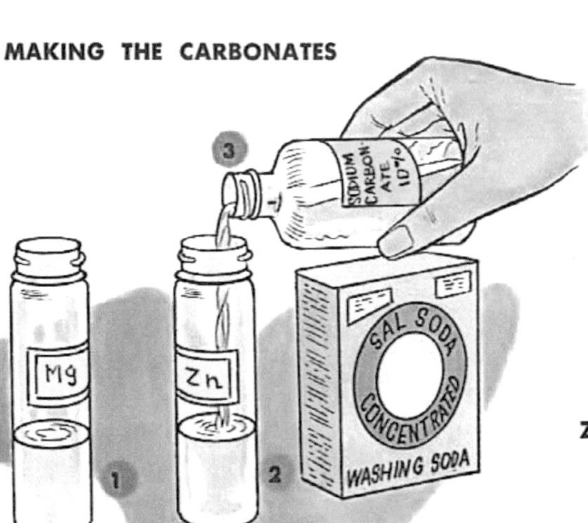

1 DISSOLVE 2 g EPSOM SALT (MAGNESIUM SULFATE, $MgSO_4 \cdot 7H_2O$) IN 20 ml WATER.

2 GET FROM HARDWARE STORE SMALL BOTTLE OF "TINNERS' FLUID." THIS IS A STRONG SOLUTION OF ZINC CHLORIDE. DILUTE 5ml OF FLUID WITH 15 ml WATER.

3 MAKE SOLUTION OF 5 g WASHING SODA (SODIUM CARBONATE) IN 50 ml WATER. ADD SOME OF THIS SOLUTION TO THE OTHER TWO. IN BOTH JARS YOU WILL GET A HEAVY WHITE PRECIPITATE. IN THE Mg JAR, THIS IS NORMAL MAGNESIUM CARBONATE ($MgCO_3$). IN Zn JAR, CO_2 IS SET FREE AND BASIC ZINC CARBONATE ($Zn(OH)_2, ZnCO_3$) RESULTS.

Zn AND Mg WITH H_2S

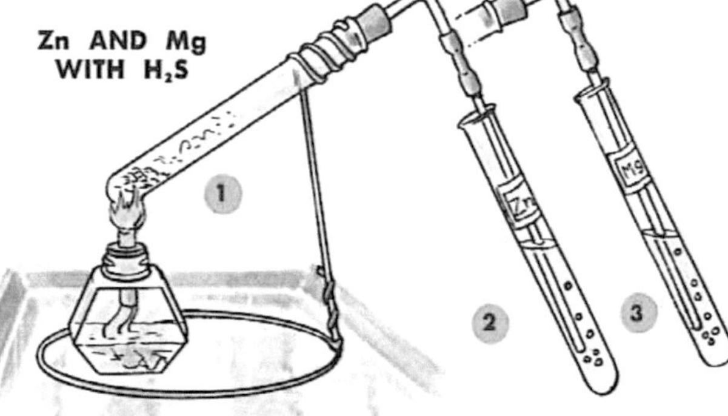

1 SET UP HYDROGEN SULFIDE APPARATUS SHOWN ON PAGE 53.

2 LEAD HYDROGEN SULFIDE (H_2S) INTO DILUTED TINNERS' FLUID ($ZnCl_2$). YOU GET A WHITE PRECIPITATE OF ZnS.

3 LEAD H_2S INTO SOLUTION OF EPSOM SALT ($MgSO_4$). HERE ALSO YOU GET WHITE PRECIPITATE. BUT NOT OF MAGNESIUM SULFIDE. THIS REACTS WITH THE WATER TO MAKE $Mg(OH)_2$.

Aluminum—in Abundance

IT IS ALMOST impossible to imagine our world without aluminum. Almost everywhere you look you see items made of this silver-white metal — from the pots in the kitchen to the airplanes flying overhead.

Although aluminum is the most abundant metal on earth, no one had ever seen it until 1825 when a Danish scientist, Hans Christian Ørsted, isolated it from aluminum chloride ($AlCl_3$). For a number of years aluminum was so expensive that it was considered in class with gold and silver. The solid aluminum cap placed on top of the Washington Monument in 1884 was first put on public display so that everyone could have a look at such a great rarity. Two

H. C. ØRSTED OF DENMARK DISCOVERED ALUMINUM IN 1825. CHARLES HALL OF THE UNITED STATES FOUND A CHEAP WAY OF PRODUCING IT IN 1886.

Al ALUMINUM Element 13. Atomic wt.: 26.98. Density: 2.70. Silver-white metal; ductile, malleable, able to take a high polish. Amphoteric. Will burn in oxygen with white flame.

MAKE A SMALL AMOUNT OF ALUMINUM POWDER BY FILING IT OFF AN OLD ALUMINUM POT. SPRINKLE IN FLAME TO MAKE SPARKS OF BURNING ALUMINUM.

GROWING ALUM CRYSTALS

1 HEAT WATER UNTIL IT IS SLIGHTLY MORE THAN LUKEWARM. STIR INTO IT POTASSIUM ALUM OR AMMONIUM ALUM UNTIL NO MORE DISSOLVES. POUR LIQUID OFF UNDISSOLVED ALUM. SET ASIDE TO COOL.

2 WHEN CRYSTALS HAVE FORMED, PICK OUT LARGEST ONES. ADD TO SOLUTION AS MUCH MORE ALUM AS IS REPRESENTED BY THE CRYSTALS YOU REMOVED. HEAT GENTLY AGAIN UNTIL ALL IS DISSOLVED. COOL.

3 POUR COOLED SOLUTION INTO NARROW GLASS. TIE THREAD TO LARGEST CRYSTAL YOU PICKED. HANG THIS IN SOLUTION FROM A PENCIL. PLACE IN QUIET SPOT. LET THE CRYSTAL GROW FOR A WEEK OR MORE.

DISSOLVING ALUMINUM

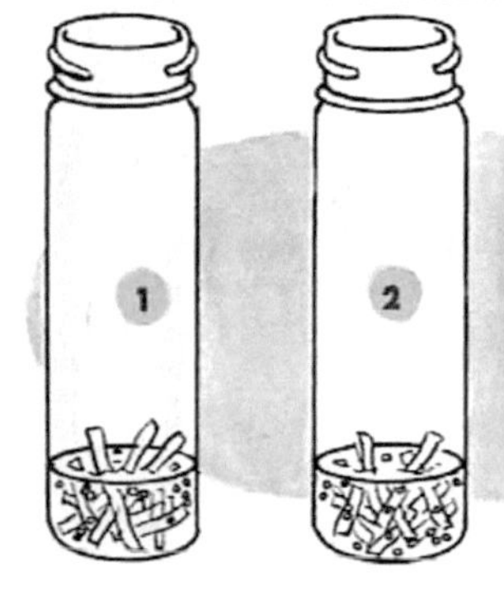

1 CUT ALUMINUM FOIL IN SMALL STRIPS. DROP THEM IN A LITTLE DILUTED HYDROCHLORIC ACID. HYDROGEN IS RELEASED; ALUMINUM CHLORIDE IS FORMED.

2 DROP STRIPS OF ALUMINUM FOIL IN 10% NaOH SOLUTION. HYDROGEN IS FREED AND SODIUM ALUMINATE ($NaAlO_2$) IS FORMED.

years later, a 22-year-old American chemist, Charles Martin Hall, invented a way of producing aluminum cheaply from aluminum oxide (Al_2O_3). Since then aluminum has become one of the most popular of all metals — mostly because of its lightness.

The mineral bauxite ($AlHO_2$, $Al(OH)_3$) is our main source of aluminum. But aluminum is also found in nature as oxide and in many complex silicates. Clay, for instance, is an aluminum silicate.

Two things about aluminum will interest you as a chemist. One is that aluminum is an "amphoteric" element, which means that it can form not only a base ($Al(OH)_3$), but also an acid ($HAlO_2$). The other is that aluminum sulfate ($Al_2(SO_4)_3$) has the ability to combine with potassium sulfate (K_2SO_4) and ammonium sulfate ($(NH_4)_2SO_4$) into beautiful cubic crystals of double salts called "alums"— $KAl(SO_4)_2 \cdot 12H_2O$ and $NH_4Al(SO_4)_2 \cdot 12H_2O$.

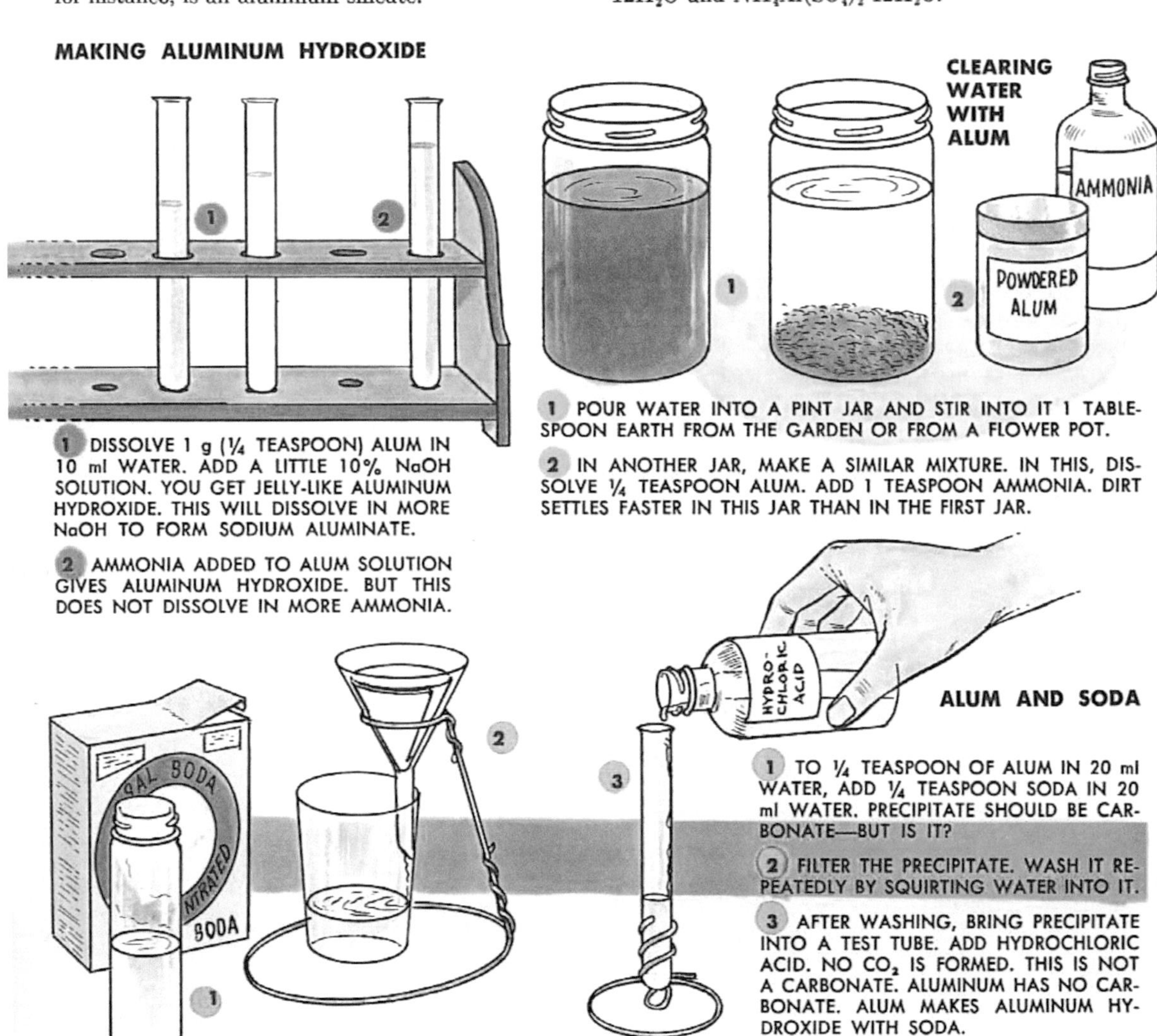

MAKING ALUMINUM HYDROXIDE

1 DISSOLVE 1 g (¼ TEASPOON) ALUM IN 10 ml WATER. ADD A LITTLE 10% NaOH SOLUTION. YOU GET JELLY-LIKE ALUMINUM HYDROXIDE. THIS WILL DISSOLVE IN MORE NaOH TO FORM SODIUM ALUMINATE.

2 AMMONIA ADDED TO ALUM SOLUTION GIVES ALUMINUM HYDROXIDE. BUT THIS DOES NOT DISSOLVE IN MORE AMMONIA.

CLEARING WATER WITH ALUM

1 POUR WATER INTO A PINT JAR AND STIR INTO IT 1 TABLESPOON EARTH FROM THE GARDEN OR FROM A FLOWER POT.

2 IN ANOTHER JAR, MAKE A SIMILAR MIXTURE. IN THIS, DISSOLVE ¼ TEASPOON ALUM. ADD 1 TEASPOON AMMONIA. DIRT SETTLES FASTER IN THIS JAR THAN IN THE FIRST JAR.

ALUM AND SODA

1 TO ¼ TEASPOON OF ALUM IN 20 ml WATER, ADD ¼ TEASPOON SODA IN 20 ml WATER. PRECIPITATE SHOULD BE CARBONATE—BUT IS IT?

2 FILTER THE PRECIPITATE. WASH IT REPEATEDLY BY SQUIRTING WATER INTO IT.

3 AFTER WASHING, BRING PRECIPITATE INTO A TEST TUBE. ADD HYDROCHLORIC ACID. NO CO_2 IS FORMED. THIS IS NOT A CARBONATE. ALUMINUM HAS NO CARBONATE. ALUM MAKES ALUMINUM HYDROXIDE WITH SODA.

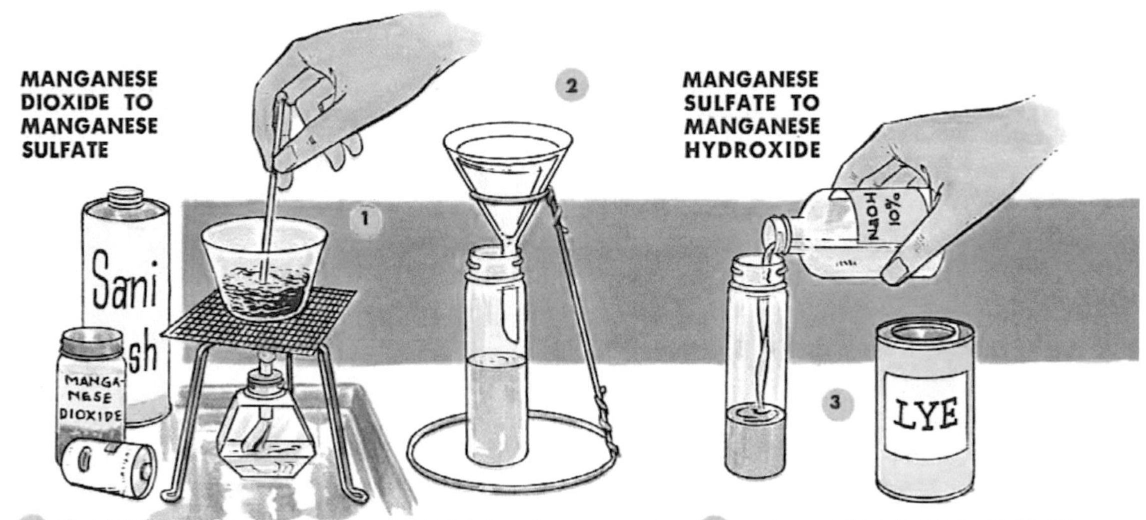

1 IN A PYREX CUSTARD CUP, MIX 2 g MANGANESE DIOXIDE, 6 g SODIUM BISULFATE, AND 10 ml WATER. HEAT MIXTURE GENTLY. IT WILL BUBBLE VIGOROUSLY BECAUSE OXYGEN IS SET FREE.

2 AFTER A FEW MINUTES, ADD 30 ml WATER. FILTER. FILTRATE CONTAINS MANGANESE SULFATE ($MnSO_4$) AND SODIUM SULFATE.

3 INTO HALF OF THE MANGANESE SULFATE SOLUTION YOU HAVE MADE, POUR 10% SOLUTION OF NaOH UNTIL NO MORE PRECIPITATE FORMS. WHITISH $Mn(OH)_2$ OXIDIZES INTO BROWN MnO(OH).

Manganese—Metal of Many Colors

METALLIC MANGANESE has no use by itself. But add up to 15 per cent of it to steel and the result is an alloy — "manganese steel" — so hard that it is suitable for machine parts that are exposed to a lot of rough wear.

The most common ore from which manganese is extracted goes under the name of "pyrolusite." This is nothing but your old friend manganese dioxide (MnO_2) which you found in your flashlight battery and have already used in a great number of your chemical experiments.

The compounds of manganese come in almost any color you can think of: black and brown, white and pink and red, violet and green. In working with these compounds, your fingers and glassware may get brown. You can get rid of this stain easily with diluted hydrochloric acid. Rinse thoroughly with water afterwards.

EXPERIMENTS WITH POTASSIUM PERMANGANATE

$KMnO_4$ WILL GIVE YOU AN IDEA OF SMALLNESS OF MOLECULE.

1 DISSOLVE ½ g POTASSIUM PERMANGANATE IN 50 ml WATER. THIS GIVES A SOLUTION OF 1 TO 100, OR 1/100.

2 DILUTE 5 ml OF THIS SOLUTION WITH 45 ml WATER. YOU NOW HAVE A SOLUTION OF 1 TO 1,000, OR 1/1,000.

3 AGAIN, 5 ml TO 45 ml WATER FOR SOLUTION 1/10,000.

4 AGAIN, 5 ml TO 45 ml WATER FOR SOLUTION 1/100,000.

5 AGAIN, 5 ml TO 45 ml WATER FOR SOLUTION 1/1,000,000. COLOR YOU STILL SEE IS CAUSED BY THE PRESENCE OF MORE THAN 600,000,000,000,000,000 MOLECULES OF $KMnO_4$.

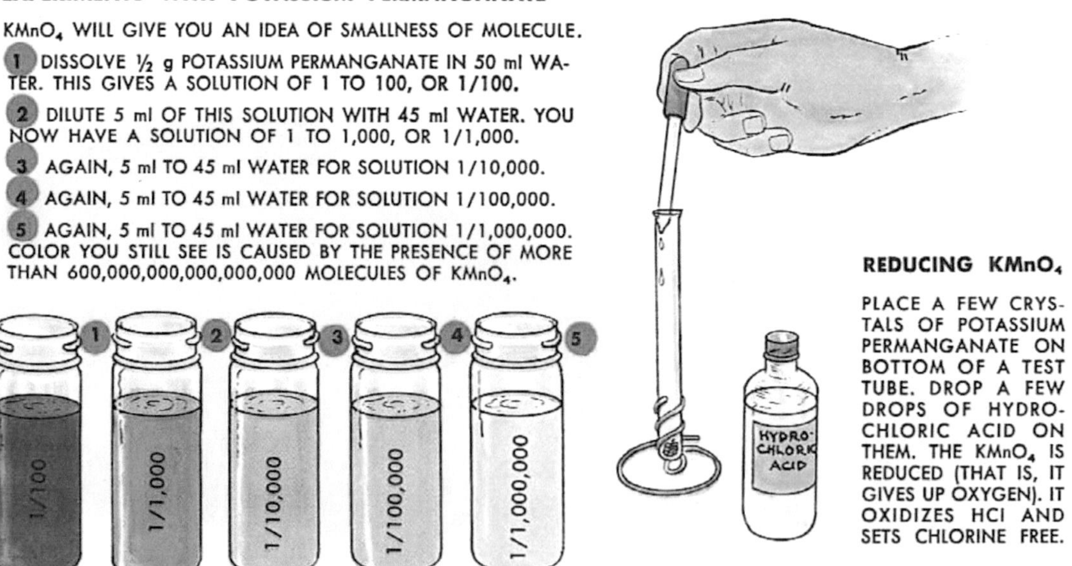

REDUCING $KMnO_4$

PLACE A FEW CRYSTALS OF POTASSIUM PERMANGANATE ON BOTTOM OF A TEST TUBE. DROP A FEW DROPS OF HYDROCHLORIC ACID ON THEM. THE $KMnO_4$ IS REDUCED (THAT IS, IT GIVES UP OXYGEN). IT OXIDIZES HCl AND SETS CHLORINE FREE.

THE EXPERIMENTS ALONG THE TOP OF THESE PAGES SHOW HOW IT IS POSSIBLE TO MOVE FROM ONE COMPOUND TO ANOTHER.

Mn	**MANGANESE**
Element 25.	

Atomic wt.: 54.94. Density: 7.44. Silvery-gray metal with reddish tinge. Reacts with water. Its compounds with oxygen range from bases to acids.

MANGANESE SULFATE TO MANGANESE CARBONATE

4 INTO SECOND HALF OF SOLUTION, POUR SOLUTION OF 4 g SODIUM CARBONATE IN 10 ml WATER. WHITE PRECIPITATE IS $MnCO_3$.

MANGANESE CARBONATE TO MANGANESE CHLORIDE

5 LET MANGANESE CARBONATE SETTLE. POUR LIQUID OFF PRECIPITATE. ADD HYDROCHLORIC ACID BY THE DROP UNTIL DISSOLVED. RESULT IS MANGANESE CHLORIDE ($MnCl_2$).

MANGANESE CHLORIDE TO MANGANESE SULFIDE

6 SET UP APPARATUS FOR MAKING HYDROGEN SULFIDE (SEE PAGE 53). POUR SOLUTION OF $MnCl_2$ INTO TEST TUBE. DILUTE IT IF NECESSARY. LEAD H_2S INTO IT. YOU GET MANGANESE SULFIDE.

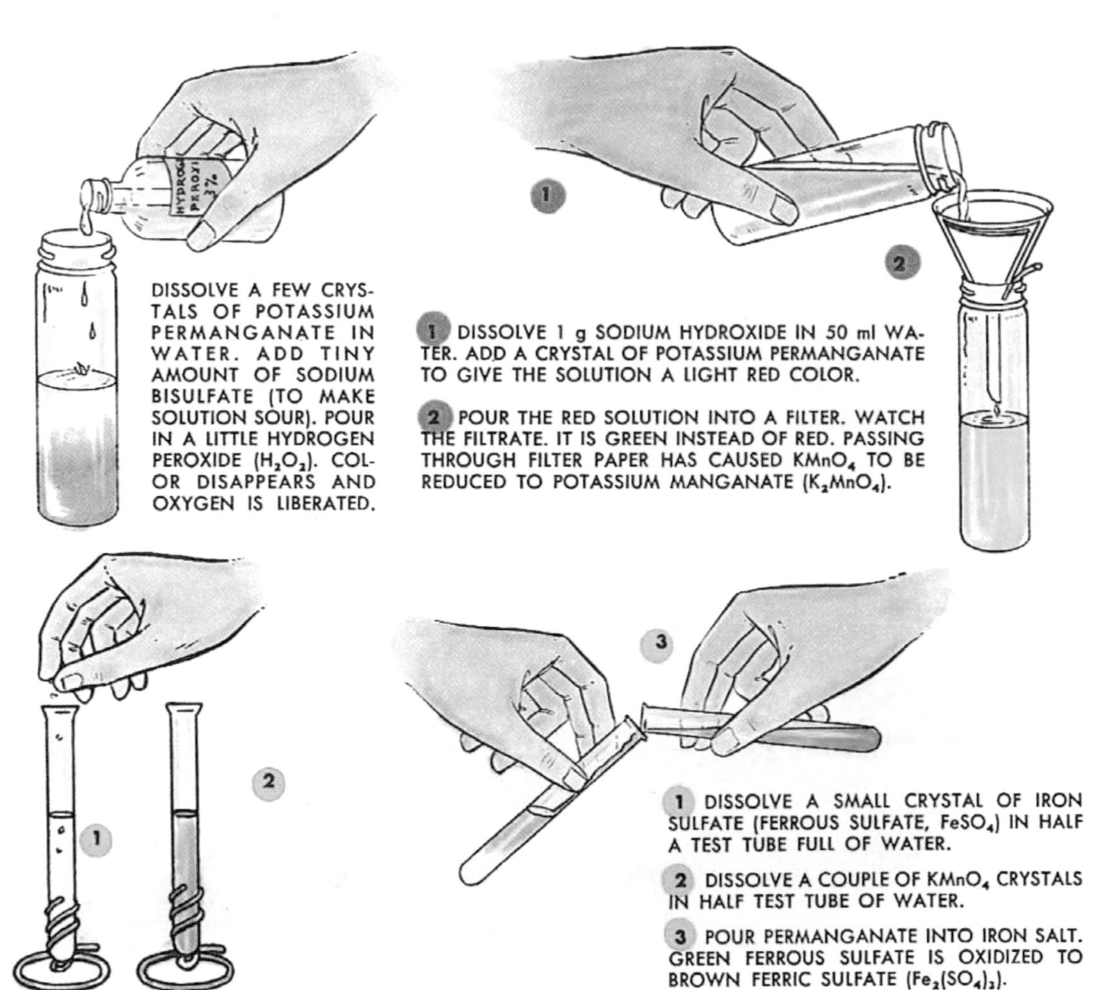

DISSOLVE A FEW CRYSTALS OF POTASSIUM PERMANGANATE IN WATER. ADD TINY AMOUNT OF SODIUM BISULFATE (TO MAKE SOLUTION SOUR). POUR IN A LITTLE HYDROGEN PEROXIDE (H_2O_2). COLOR DISAPPEARS AND OXYGEN IS LIBERATED.

1 DISSOLVE 1 g SODIUM HYDROXIDE IN 50 ml WATER. ADD A CRYSTAL OF POTASSIUM PERMANGANATE TO GIVE THE SOLUTION A LIGHT RED COLOR.

2 POUR THE RED SOLUTION INTO A FILTER. WATCH THE FILTRATE. IT IS GREEN INSTEAD OF RED. PASSING THROUGH FILTER PAPER HAS CAUSED $KMnO_4$ TO BE REDUCED TO POTASSIUM MANGANATE (K_2MnO_4).

1 DISSOLVE A SMALL CRYSTAL OF IRON SULFATE (FERROUS SULFATE, $FeSO_4$) IN HALF A TEST TUBE FULL OF WATER.

2 DISSOLVE A COUPLE OF $KMnO_4$ CRYSTALS IN HALF TEST TUBE OF WATER.

3 POUR PERMANGANATE INTO IRON SALT. GREEN FERROUS SULFATE IS OXIDIZED TO BROWN FERRIC SULFATE ($Fe_2(SO_4)_3$).

We Live in an Age of Iron

IRON METAL has the peculiar quality of being magnetic — that is, of being attracted and influenced by a force called magnetism. If you should walk around your home and touch different things with a magnet, you would be surprised at the large number of them that would prove to contain iron. They would range in size from the car in the garage and the refrigerator and stove in the kitchen to the nails in the walls and the needles and pins in your mother's sewing box.

The moment you step outdoors and look around, you will be even more amazed. Skyscrapers and

TWO KINDS OF IRON COMPOUNDS

IRON FORMS TWO KINDS OF COMPOUNDS. IN FERROUS SALTS, EACH IRON ATOM HAS REPLACED TWO HYDROGEN ATOMS. IN FERRIC SALTS, EACH IRON ATOM HAS REPLACED THREE HYDROGEN ATOMS. GREEN FERROUS SALTS EASILY OXIDIZE INTO RED-BROWN FERRIC SALTS.

RUSTING OF IRON

MOISTEN A WAD OF FINE STEEL WOOL WITH VINEGAR (TO SPEED UP ACTION). WEDGE IT IN BOTTOM OF A GLASS. INVERT GLASS IN PIE PLATE OF WATER. IN A FEW DAYS, WATER HAS RISEN IN GLASS. IRON HAS REACTED WITH OXYGEN AND MOISTURE TO FORM RUST—$(Fe_2O_3)_2 \cdot 3H_2O$.

MAKING A FERROUS SALT

1 POUR HYDROCHLORIC ACID OVER STEEL WOOL. HYDROGEN IS SET FREE AS STEEL WOOL DISSOLVES. FILTER THE SOLUTION.

2 LIGHT-GREEN FILTRATE CONTAINS FERROUS CHLORIDE ($FeCl_2$).

MAKING A FERRIC SALT

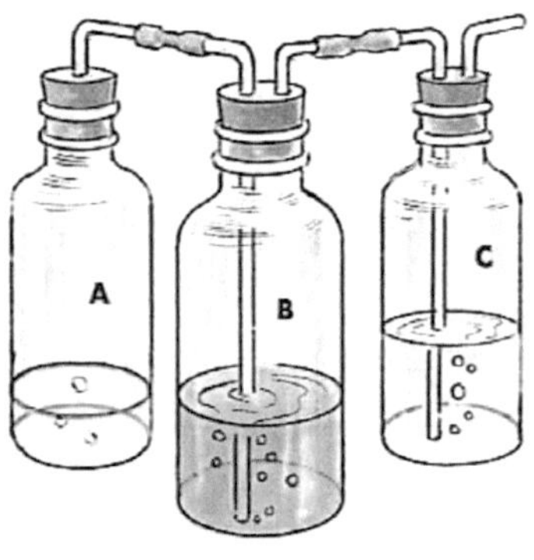

SET UP APPARATUS FOR MAKING CHLORINE (SEE PAGE 35). INTO BOTTLE **B** POUR FERROUS CHLORIDE SOLUTION YOU HAVE JUST MADE. THE CHLORINE TURNS THE GREEN FERROUS CHLORIDE ($FeCl_2$) INTO A BROWN FERRIC CHLORIDE ($FeCl_3$).

TEST FOR IRON SALTS

1 IN ONE TEST TUBE, DILUTE SOME FERRIC CHLORIDE SOLUTION WITH WATER.

2 IN ANOTHER, DILUTE SOME OF THE FERROUS CHLORIDE SOLUTION WITH WATER.

3 TO EACH, ADD A FEW DROPS OF SOLUTION OF 1/4 TEASPOON POTASSIUM FERROCYANIDE IN 50 ml WATER. FERRIC SALT MAKES A DEEP BLUE PRECIPITATE OF PRUSSIAN BLUE. FERROUS SALT MAKES LIGHT BLUE PRECIPITATE.

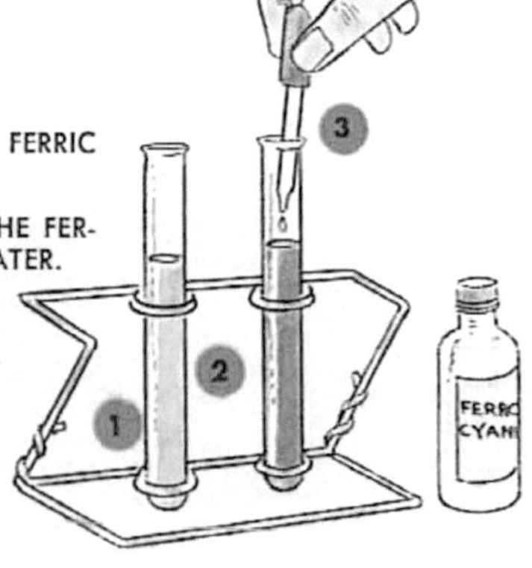

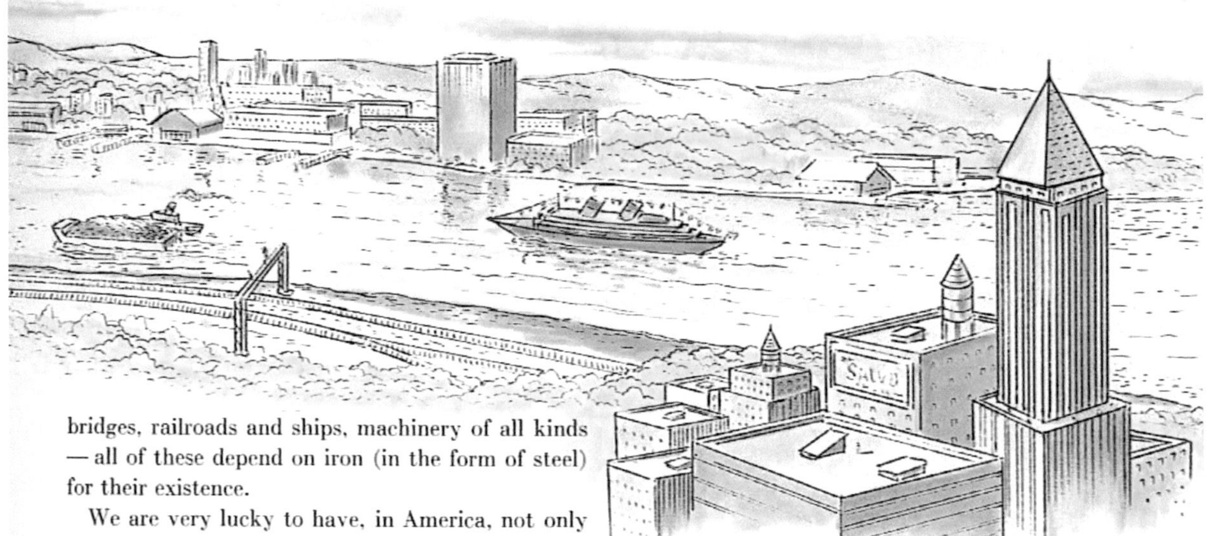

bridges, railroads and ships, machinery of all kinds — all of these depend on iron (in the form of steel) for their existence.

We are very lucky to have, in America, not only large amounts of iron ore but also large amounts of the coal from which to make the coke that goes into iron production.

The iron is driven out of its ore (mostly Fe_2O_3) in huge furnaces. Each furnace can make as much as 1,000 tons of iron at one time from 2,000 tons of ore, 1,000 tons of coke, and 500 tons of limestone. A blast of hot air is forced through the mixture. The coke burns with great heat to carbon dioxide. This, with more coke, forms carbon monoxide, and this, in turn, reduces the iron oxide to metallic iron. In chemical language, this is what happens:

$$C + O_2 \rightarrow CO_2 \text{ plus heat}$$
$$CO_2 + C \rightarrow 2CO$$
$$Fe_2O_3 + 3CO \rightarrow 2Fe + 3CO_2$$

At the same time, the limestone combines with various impurities to form a glass-like compound called "slag." This is removed when the white-hot iron is poured out into moulds and cooled into bars of "pig iron."

The pig iron is brittle because it contains close to 5 per cent carbon. To turn it into steel, the carbon must be burned out until only from .5 to 1.5 per cent remains. This is done either by the Bessemer process (named for an Englishman, Henry Bessemer) or by the open-hearth process. The finished steel is molded into "ingots" and shipped to manufacturing plants all over the country.

In chemical experiments, the most commonly used iron compound is the iron sulfate (ferrous sulfate, $FeSO_4 \cdot 7H_2O$) — also called "green vitriol" and "copperas." Don't let the last name mislead you— it has nothing to do with copper but comes from an old French word, *couperose*.

FROM FERROUS SALT TO FERRIC

1 DISSOLVE ¼ TEASPOON FERROUS SULFATE IN 50 ml WATER. ADD A FEW CRYSTALS OF SODIUM BISULFATE TO KEEP THE SOLUTION SOUR.

2 ADD HYDROGEN PEROXIDE SOLUTION. LIGHT-GREEN FERROUS SULFATE SOLUTION TURNS REDDISH-BROWN. H_2O_2 HAS OXIDIZED $FeSO_4$ TO FERRIC SULFATE ($Fe_2(SO_4)_3$).

IRON HYDROXIDES AND CARBONATE

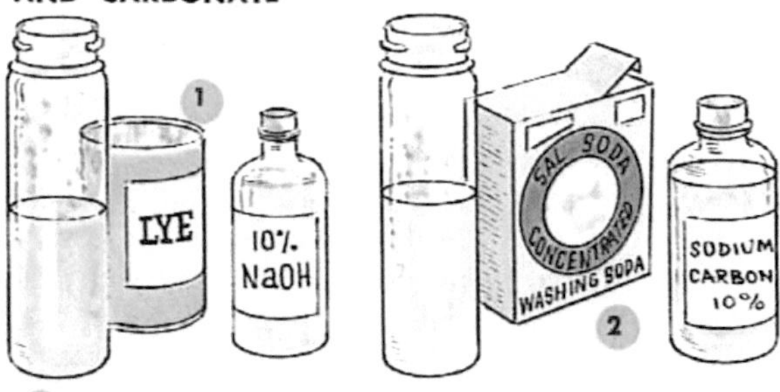

1 TO SOLUTION OF ¼ TEASPOON FERROUS SULFATE IN 50 ml WATER, ADD SOLUTION OF SODIUM HYDROXIDE. PURE FERROUS HYDROXIDE IS WHITE. BECAUSE OF IMPURITIES, YOU GET DIRTY-GREEN PRECIPITATE OF $Fe(OH)_2$, SOON OXIDIZING TO BROWN FERRIC HYDROXIDE.

2 TO ANOTHER PORTION OF FERROUS SULFATE SOLUTION ADD SODIUM CARBONATE SOLUTION. PURE CARBONATE MADE WITH NO OXYGEN PRESENT IS WHITE— BUT YOU GET MUDDY, WHITISH-GREEN PRECIPITATE OF FERROUS CARBONATE, EVENTUALLY TURNING INTO FERRIC HYDROXIDE.

THE GREEK AND TROJAN WARRIORS FIGHTING BEFORE THE GATES OF TROY USED SWORDS AND SHIELDS OF BRONZE—AN ALLOY MADE UP OF COPPER AND TIN.

Copper—Yesterday, Today

COPPER IS ONE of the few metals found free in nature. That is why it was used long before historic times for weapons and utensils. The main trouble with it was its softness. This was remedied when some early coppersmith discovered that copper and tin (also found free in nature) melted together formed an alloy that was much harder than either of the two metals. This alloy gave its name to more than two thousand years of human history — the period called the "Bronze Age."

A great number of weapons from the Bronze Age have been found in Greece. When they were dug out of the ground, they were covered with a green "rust." This deposit was called verdigris — literally "green of Greece" (from old French, *vert de Grèce*). It consists of basic cupric carbonate — the same compound you will see on a bronze statue or a copper-clad church spire exposed to wind and weather.

Copper became especially valuable less than a hundred years ago when a satisfactory method for producing a steady flow of electricity was invented. After silver, copper is the best conductor of electricity. Today, the most important use for copper is for electrical purposes. It serves to bring the current from the place where it is produced to the place where it is to be used (although, within recent years, some aluminum has taken its place for high-tension wires). You will find copper in the wiring in your own home and in every electrical gadget you use.

Copper makes two kinds of salts. In cuprous salts, one copper atom has taken the place of one hydrogen atom; in cupric salts, one copper atom has taken the place of two hydrogen atoms. Cuprous salts (such as cuprous chloride, $CuCl$) are colorless, while cupric salts (such as cupric sulfate, $CuSO_4 \cdot 5H_2O$) are bright blue in color.

MOST IMPORTANT USE OF COPPER TODAY IS FOR ELECTRIC WIRING.

MAKING COPPER COMPOUNDS

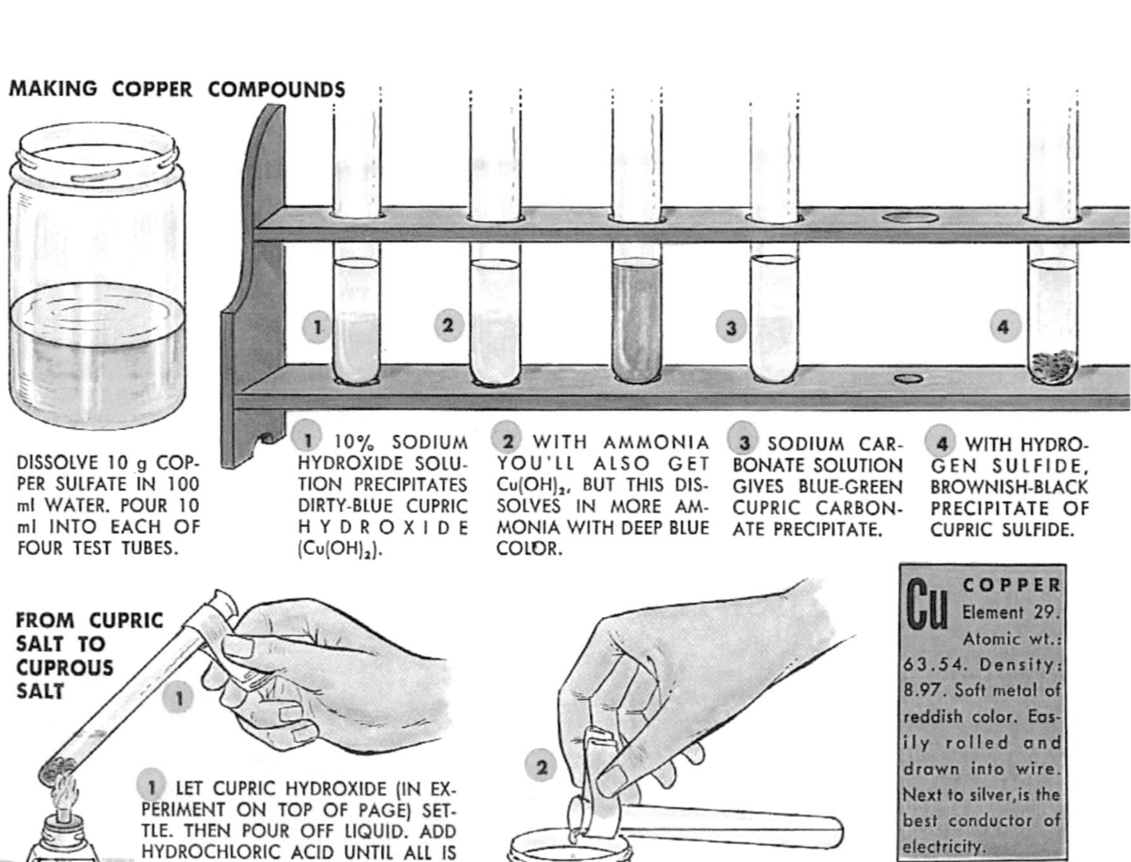

DISSOLVE 10 g COPPER SULFATE IN 100 ml WATER. POUR 10 ml INTO EACH OF FOUR TEST TUBES.

1 10% SODIUM HYDROXIDE SOLUTION PRECIPITATES DIRTY-BLUE CUPRIC HYDROXIDE ($Cu(OH)_2$).

2 WITH AMMONIA YOU'LL ALSO GET $Cu(OH)_2$, BUT THIS DISSOLVES IN MORE AMMONIA WITH DEEP BLUE COLOR.

3 SODIUM CARBONATE SOLUTION GIVES BLUE-GREEN CUPRIC CARBONATE PRECIPITATE.

4 WITH HYDROGEN SULFIDE, BROWNISH-BLACK PRECIPITATE OF CUPRIC SULFIDE.

Cu COPPER Element 29. Atomic wt.: 63.54. Density: 8.97. Soft metal of reddish color. Easily rolled and drawn into wire. Next to silver, is the best conductor of electricity.

FROM CUPRIC SALT TO CUPROUS SALT

1 LET CUPRIC HYDROXIDE (IN EXPERIMENT ON TOP OF PAGE) SETTLE. THEN POUR OFF LIQUID. ADD HYDROCHLORIC ACID UNTIL ALL IS DISSOLVED. ADD SMALL PIECES OF COPPER WIRE. HEAT TO BOILING.

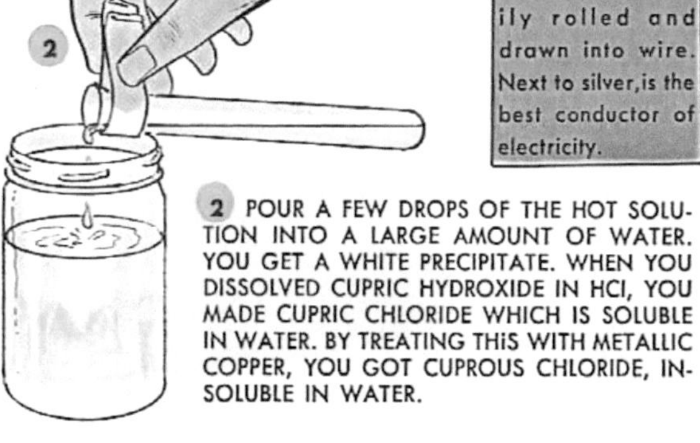

2 POUR A FEW DROPS OF THE HOT SOLUTION INTO A LARGE AMOUNT OF WATER. YOU GET A WHITE PRECIPITATE. WHEN YOU DISSOLVED CUPRIC HYDROXIDE IN HCl, YOU MADE CUPRIC CHLORIDE WHICH IS SOLUBLE IN WATER. BY TREATING THIS WITH METALLIC COPPER, YOU GOT CUPROUS CHLORIDE, INSOLUBLE IN WATER.

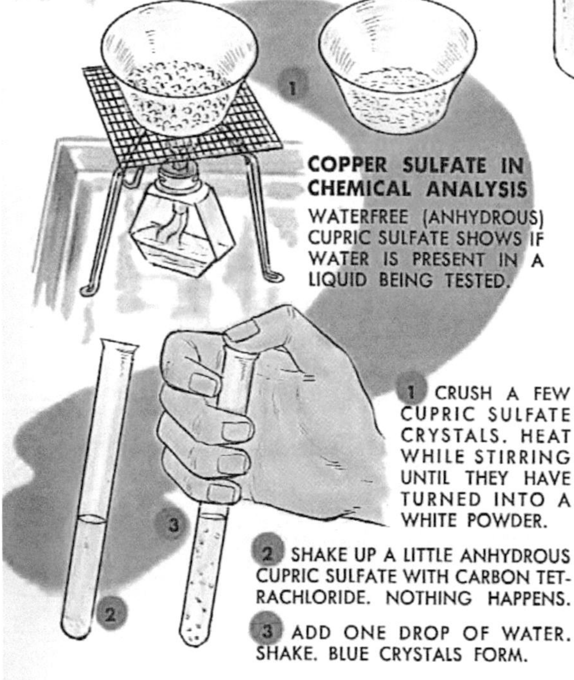

COPPER SULFATE IN CHEMICAL ANALYSIS

WATERFREE (ANHYDROUS) CUPRIC SULFATE SHOWS IF WATER IS PRESENT IN A LIQUID BEING TESTED.

1 CRUSH A FEW CUPRIC SULFATE CRYSTALS. HEAT WHILE STIRRING UNTIL THEY HAVE TURNED INTO A WHITE POWDER.

2 SHAKE UP A LITTLE ANHYDROUS CUPRIC SULFATE WITH CARBON TETRACHLORIDE. NOTHING HAPPENS.

3 ADD ONE DROP OF WATER. SHAKE. BLUE CRYSTALS FORM.

REPLACING COPPER WITH IRON

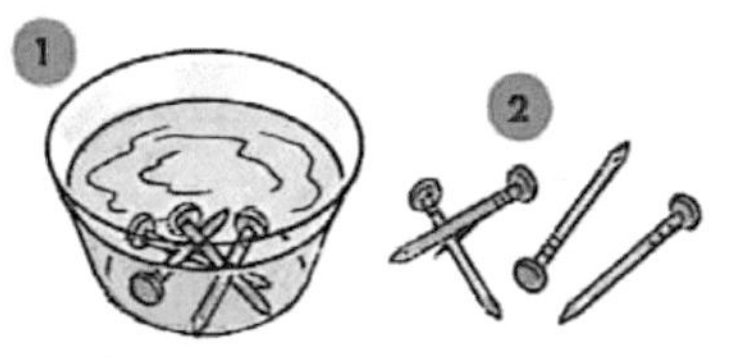

1 DROP SEVERAL CLEAN NAILS INTO A SOLUTION OF COPPER SULFATE. LEAVE FOR HALF AN HOUR.

2 NAILS ARE NOW COATED WITH METALLIC COPPER AND THE SOLUTION CONTAINS FERROUS SULFATE.

METALS CAN BE ARRANGED IN A REPLACEMENT SERIES. ANY METAL IN THE SERIES WILL DRIVE OUT ANOTHER METAL BELOW IT AND TAKE ITS PLACE IN THE SALT.

REPLACEMENT SERIES

POTASSIUM
SODIUM
CALCIUM
MAGNESIUM
ALUMINUM
ZINC
CHROMIUM
IRON
NICKEL
TIN
LEAD
COPPER
MERCURY
SILVER
PLATINUM
GOLD

Silver—One of the "Noble" Metals

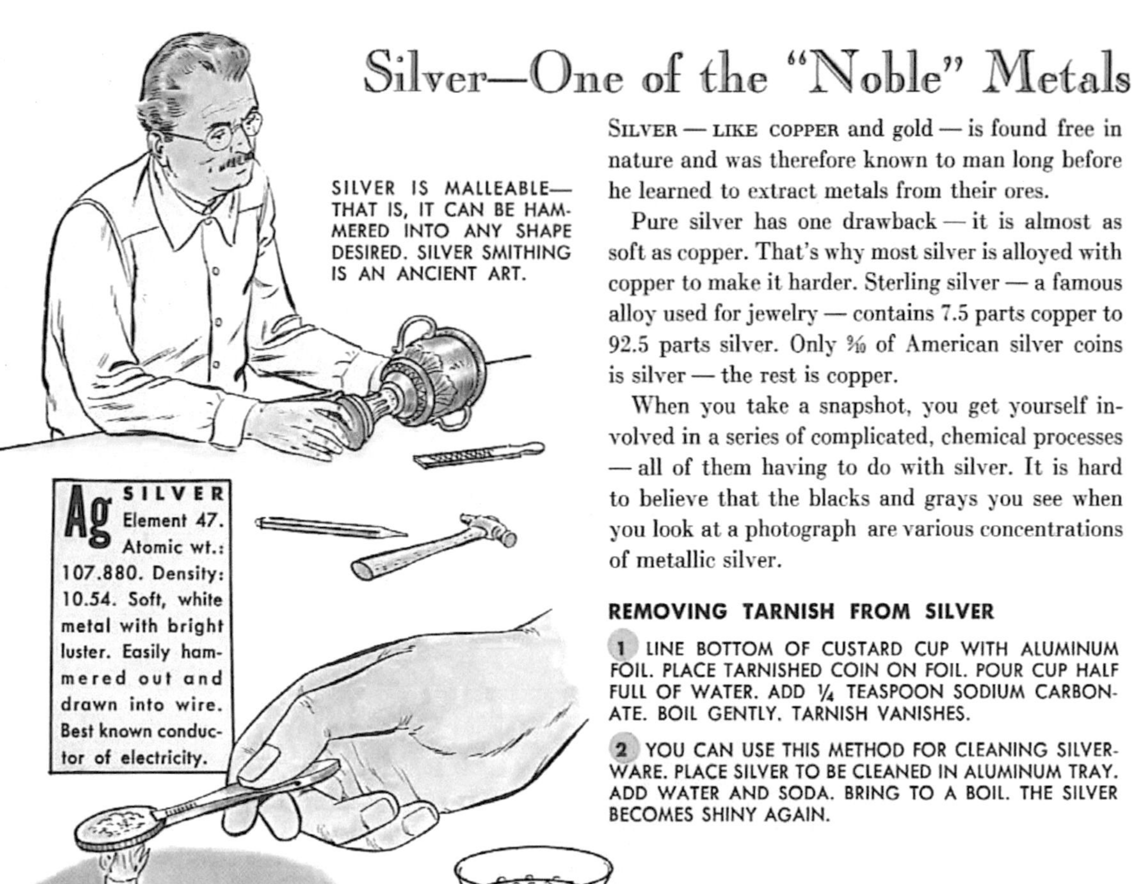

SILVER — LIKE COPPER and gold — is found free in nature and was therefore known to man long before he learned to extract metals from their ores.

Pure silver has one drawback — it is almost as soft as copper. That's why most silver is alloyed with copper to make it harder. Sterling silver — a famous alloy used for jewelry — contains 7.5 parts copper to 92.5 parts silver. Only 9⁄10 of American silver coins is silver — the rest is copper.

When you take a snapshot, you get yourself involved in a series of complicated, chemical processes — all of them having to do with silver. It is hard to believe that the blacks and grays you see when you look at a photograph are various concentrations of metallic silver.

REMOVING TARNISH FROM SILVER

1 LINE BOTTOM OF CUSTARD CUP WITH ALUMINUM FOIL. PLACE TARNISHED COIN ON FOIL. POUR CUP HALF FULL OF WATER. ADD 1⁄4 TEASPOON SODIUM CARBONATE. BOIL GENTLY. TARNISH VANISHES.

2 YOU CAN USE THIS METHOD FOR CLEANING SILVERWARE. PLACE SILVER TO BE CLEANED IN ALUMINUM TRAY. ADD WATER AND SODA. BRING TO A BOIL. THE SILVER BECOMES SHINY AGAIN.

TARNISHED SILVER
SILVER TARNISHES WHEN IT IS EXPOSED TO SULFUR. PLACE A FEW CRYSTALS OF SODIUM THIOSULFATE ("HYPO") ON A SILVER COIN. HEAT UNTIL HYPO MELTS. WASH. HYPO HAS LEFT STAIN OF BROWN-BLACK SILVER SULFIDE.

SILVER COMPOUNDS

GET 5 g SILVER NITRATE IN YOUR LOCAL DRUG STORE. DISSOLVE IN 50 ml WATER.

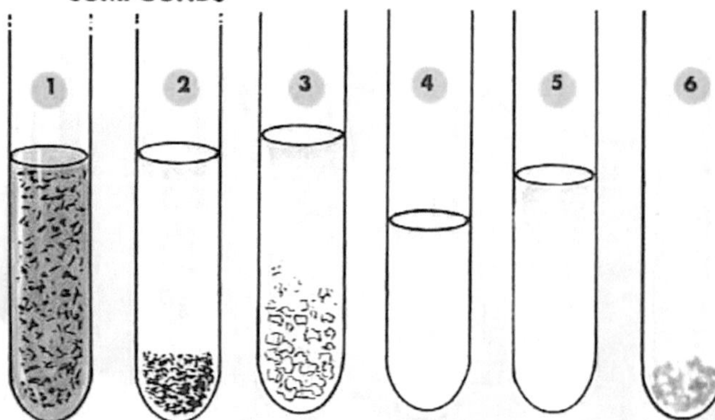

1 TO 5 ml SILVER NITRATE ($AgNO_3$) SOLUTION, ADD SODIUM HYDROXIDE SOLUTION. YOU GET DARK-BROWN PRECIPITATE—NOT OF HYDROXIDE, BUT OF SILVER OXIDE.

2 TO 5 ml $AgNO_3$ SOLUTION, ADD AMMONIA. PRECIPITATE OF SILVER OXIDE DISSOLVES WHEN YOU ADD MORE AMMONIA.

3 TO 5 ml $AgNO_3$ SOLUTION, ADD TABLE SALT ($NaCl$) SOLUTION. CHEESELIKE PRECIPITATE IS SILVER CHLORIDE ($AgCl$).

4 TO PART OF $AgCl$ PRECIPITATE, ADD AMMONIA. SILVER CHLORIDE DISSOLVES.

5 TO ANOTHER PART OF $AgCl$, ADD SODIUM THIOSULFATE SOLUTION. $AgCl$ DISSOLVES.

6 PLACE REMAINING $AgCl$ IN THE SUN. IT TURNS VIOLET FROM METALLIC SILVER.

In making a photographic film, the manufacturer spreads an emulsion of gelatin that contains silver bromide ($AgBr$) over a transparent sheet of cellulose acetate. When the silver bromide is exposed to light, a certain amount of it gives up metallic silver ($AgBr \rightarrow Ag + Br$). More of this silver is brought out in the developing bath. When fully developed, the film is placed in a fixing bath which removes all unexposed silver bromide. After washing and drying, you have a photographic negative in which the white parts you photographed appear black and the black parts appear white.

To make a natural-looking picture, you place the negative on a piece of photographic paper and go through a similar procedure, as above, of exposing, developing, fixing, washing, and drying.

MORE THAN 150 TONS OF SILVER ARE USED EACH YEAR IN MAKING FILM FOR THE MOVIES.

PHOTOGRAPHY INVOLVES A WHOLE SERIES OF CHEMICAL PROCESSES.

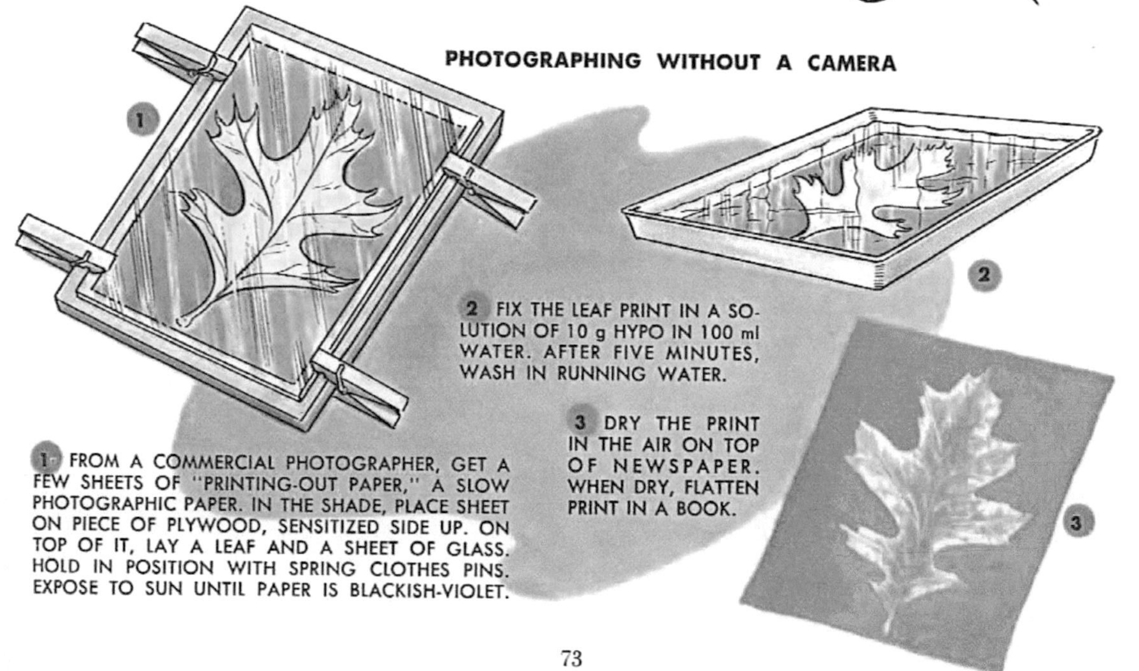

PHOTOGRAPHING WITHOUT A CAMERA

1 FROM A COMMERCIAL PHOTOGRAPHER, GET A FEW SHEETS OF "PRINTING-OUT PAPER," A SLOW PHOTOGRAPHIC PAPER. IN THE SHADE, PLACE SHEET ON PIECE OF PLYWOOD, SENSITIZED SIDE UP. ON TOP OF IT, LAY A LEAF AND A SHEET OF GLASS. HOLD IN POSITION WITH SPRING CLOTHES PINS. EXPOSE TO SUN UNTIL PAPER IS BLACKISH-VIOLET.

2 FIX THE LEAF PRINT IN A SOLUTION OF 10 g HYPO IN 100 ml WATER. AFTER FIVE MINUTES, WASH IN RUNNING WATER.

3 DRY THE PRINT IN THE AIR ON TOP OF NEWSPAPER. WHEN DRY, FLATTEN PRINT IN A BOOK.

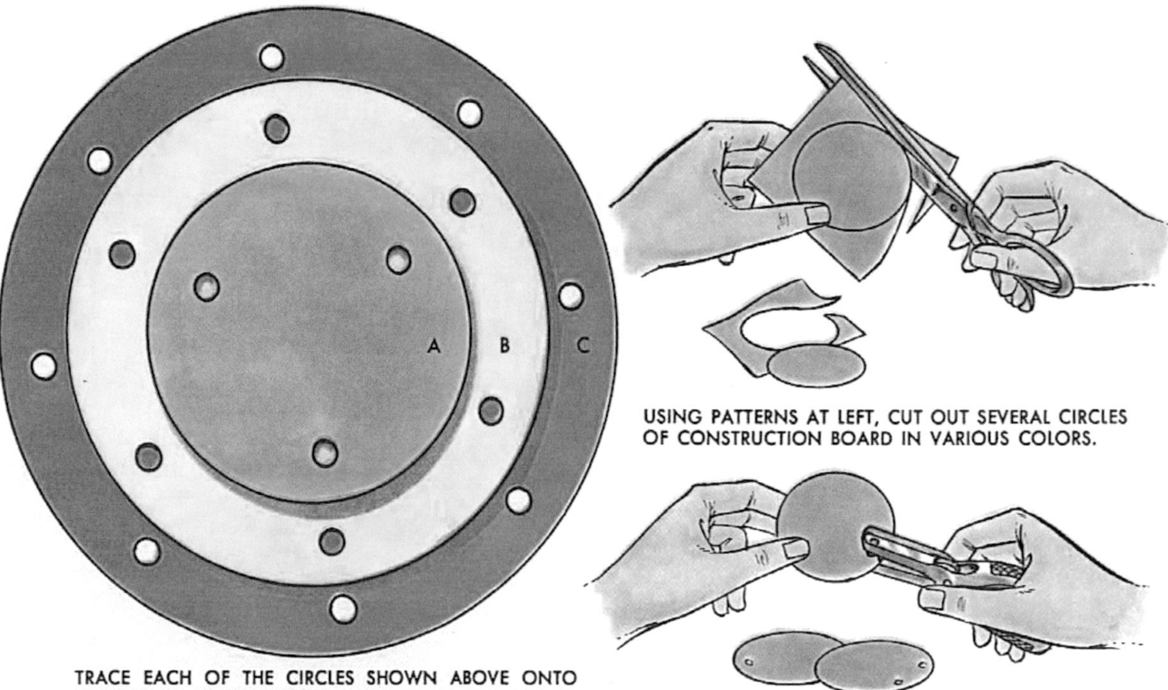

TRACE EACH OF THE CIRCLES SHOWN ABOVE ONTO CARDBOARD. PUNCH HOLES AS INDICATED. USE AS PATTERNS FOR CUTTING CIRCLES OF CONSTRUCTION BOARD.

USING PATTERNS AT LEFT, CUT OUT SEVERAL CIRCLES OF CONSTRUCTION BOARD IN VARIOUS COLORS.

PUNCH THE HOLES NECESSARY TO INDICATE VALENCES.

Valences and Formulas

As you have studied the chemical formulas in the text, you will have noticed that one atom of hydrogen combines with one atom of chlorine (HCl), two hydrogen atoms with one atom of oxygen (H_2O), and three hydrogen atoms with one atom of nitrogen (NH_3).

The capacity of one atom to hold on to other atoms is called its valence (from Latin *valentia*, strength). No atom has a lower valence than the hydrogen atom, so we use hydrogen as our starting point and give it a valence of 1. Two hydrogen atoms combine with one oxygen atom — that gives oxygen a valence of 2. Nitrogen has a valence of 3. Two oxygen atoms combine with one carbon atom to make CO_2. Carbon has a valence of 4.

The chart on page 75 shows some of the common

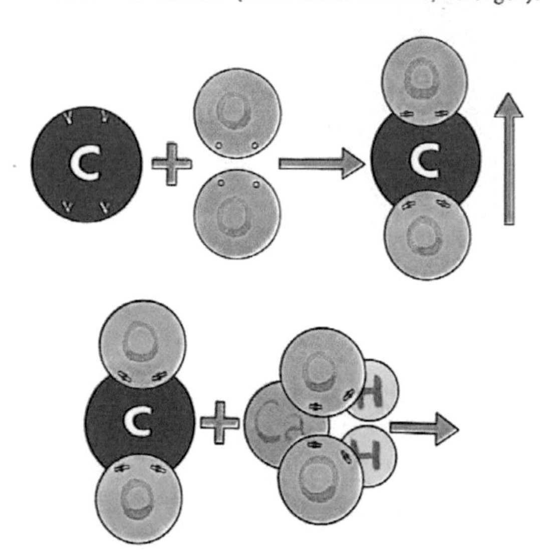

THESE DIAGRAMS SHOW WHAT HAPPENS WHEN YOU BURN CARBON AND TEST FOR CO_2. ONE CARBON ATOM (WITH FOUR POSITIVE VALENCES) COMBINES WITH TWO ATOMS OF OXYGEN (EACH WITH TWO NEGATIVE VALENCES) TO FORM ONE MOLECULE OF CO_2 (ARROW POINTING UP INDICATES THAT THIS IS A GAS). ONE MOLECULE CARBON DIOXIDE COMBINES WITH ONE MOLECULE CALCIUM HYDROXIDE TO FORM ONE MOLECULE OF CALCIUM CARBONATE (ARROW POINTING DOWN INDICATES THAT IT IS A PRECIPITATE) AND ONE MOLECULE OF WATER.

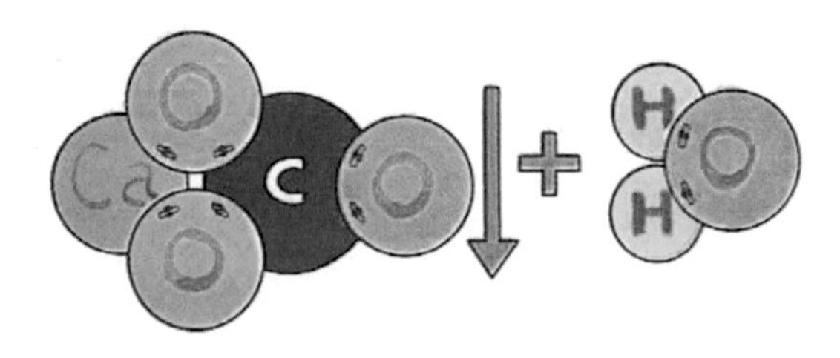

PUT ¼" BRASS CLIPS IN HOLES SHOWING POSITIVE VALENCES. HOLD THEM IN PLACE WITH SCOTCH TAPE.

WRITE THE NAMES OF THE ELEMENTS WITH CRAYONS.

SOME COMMON VALENCES

Positive Valences			Negative Valences		
Item	Valence	Circle	Item	Valence	Circle
Ag	+1	A	Cl	−1	A
Al	+3	C	I	−1	A
B	+3	C	N	−3	B
C	+4	C	O	−2	B
Ca	+2	B	S	−2	C
Cu	+1+2	B			
Fe	+2+3	B			
H	+1	A			
K	+1	A			
Mg	+2	B			
Mn	+2	B	CO_3	−2	B
Na	+1	A	NO_3	−1	A
S	+4+6	C	OH	−1	A
Si	+4	C	SO_3	−2	B
NH_4	+1	A	SO_4	−2	B

valences for making up formulas. Most of the items are elements, but some of them are "radicals" — that is, groups of atoms that hang together in chemical reactions, such as the ammonium radical (NH_4) that behaves as a metal, and the sulfate radical (SO_4) that goes into the making of salts.

Notice that some valences have plus (+) signs, others have minus (—) signs. When you make up the formula for a compound, there must be the same number of pluses and minuses. Hydrogen with one plus (H+) and oxygen with two minuses (O−−) would not fit together — you need H_2 to combine with O. Similarly, C with + 4 (C++++) takes two O, each with —2 (O−−), in order to balance.

To get a clear understanding of chemical formulas, make yourself a set of atom models as shown on these pages. With these models you will be able to figure out how compounds are made up and what happens in the various chemical reactions you will cause in your experiments.

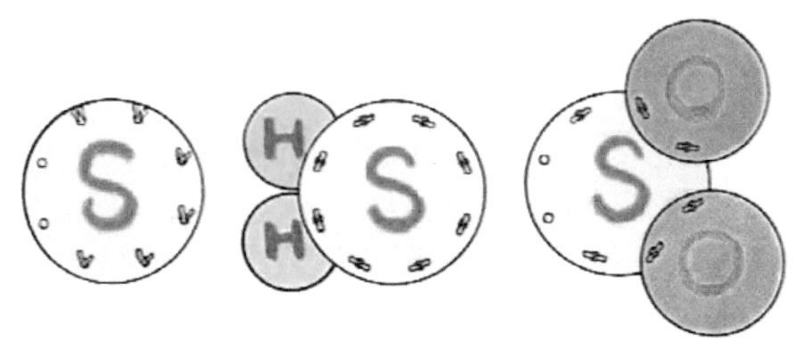

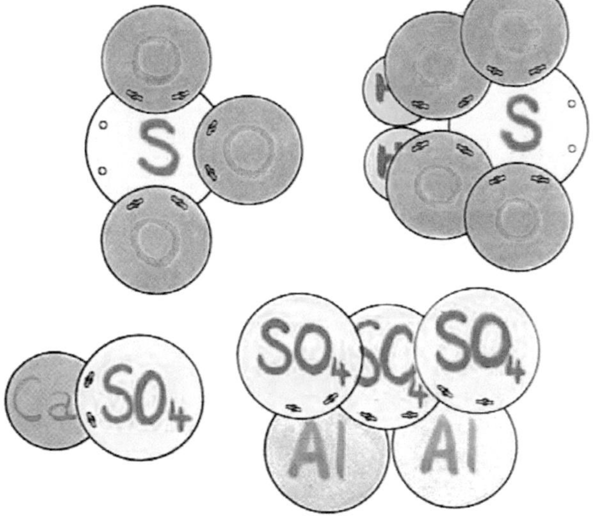

SULFUR HAS SEVERAL VALENCES. IT HAS A VALENCE OF −2 IN H_2S, OF +4 IN SO_2, AND OF +6 IN SO_3 AND IN SULFURIC ACID (H_2SO_4). IN MAKING THE CIRCLE FOR SULFUR, YOU CAN SHOW THESE VALENCES WITH TWO EMPTY HOLES AND SIX BRASS CLIPS.

INSTEAD OF USING ONE SULFUR CIRCLE AND FOUR OXYGEN CIRCLES TO INDICATE A SULFATE, YOU CAN MAKE UP A SINGLE CIRCLE TO STAND FOR THE SULFATE RADICAL (SO_4) WITH TWO NEGATIVE VALENCES.

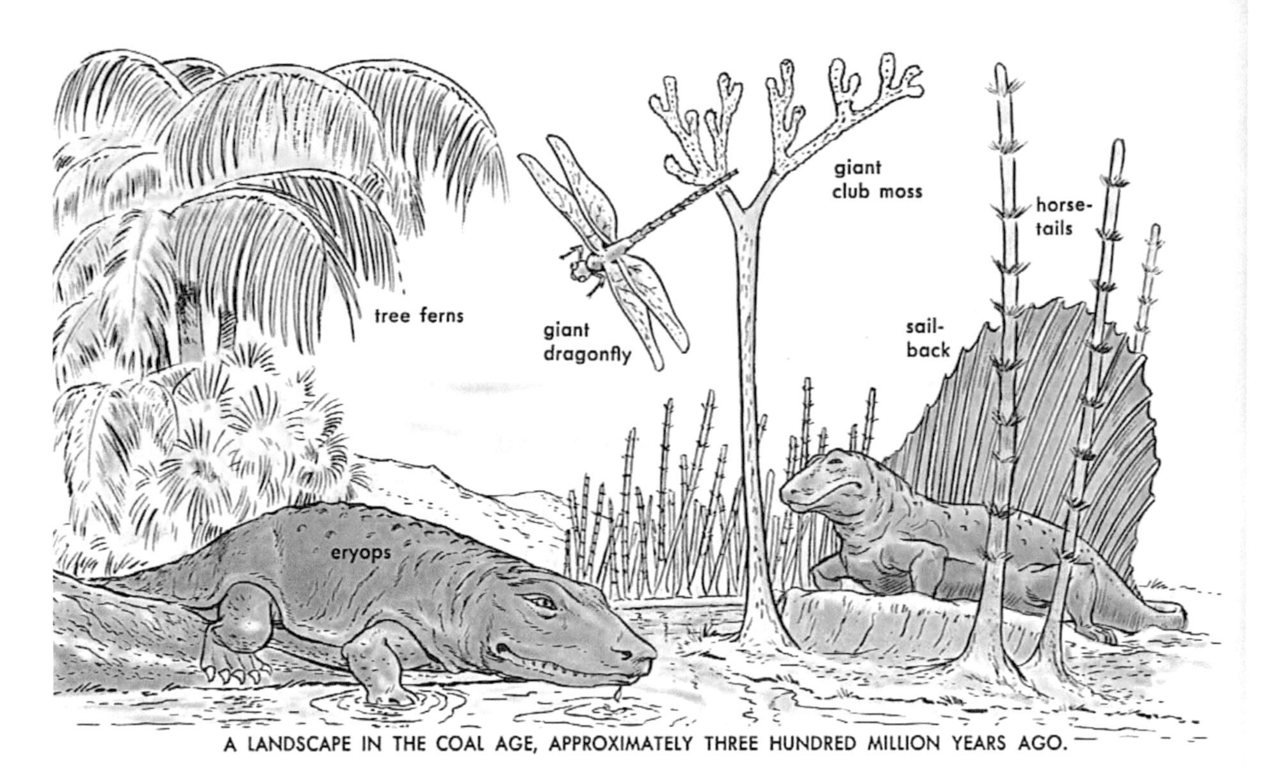

A LANDSCAPE IN THE COAL AGE, APPROXIMATELY THREE HUNDRED MILLION YEARS AGO.

Carbon—Element of a Million Compounds

TO THE OLD ROMANS, *carbo* meant coal — a black rock that would burn. To the modern chemist, carbon is an element found in all living things — plants and animals — and in many dead things. It is hidden in the whitest sugar and the reddest rose and the greenest apple, in hundreds of thousands of compounds produced by nature and in many thousands more created in the laboratory.

The soot from a smoking candle is almost pure carbon. So is also the graphite that forms the "lead" of your pencil and the diamond in the jeweler's window. The coal that we use for fuel contains from 80 to 90 per cent carbon — the other 10 to 20 per cent is made up of various substances from which a great number of important and valuable chemical compounds are made.

All the coal we mine deep underground today is made up of the remains of plants that grew around three hundred million years ago — huge tree ferns, giant club mosses and horsetails. They thrived in the hot, humid climate, died and tumbled to the ground. During the ages they were covered by other dead trees and by layers upon layers of mud. Eventually, pressure and heat turned them into coal.

PRESSURE AND HEAT TURNED TREES AND OTHER PLANTS INTO THE COAL WE USE TODAY.

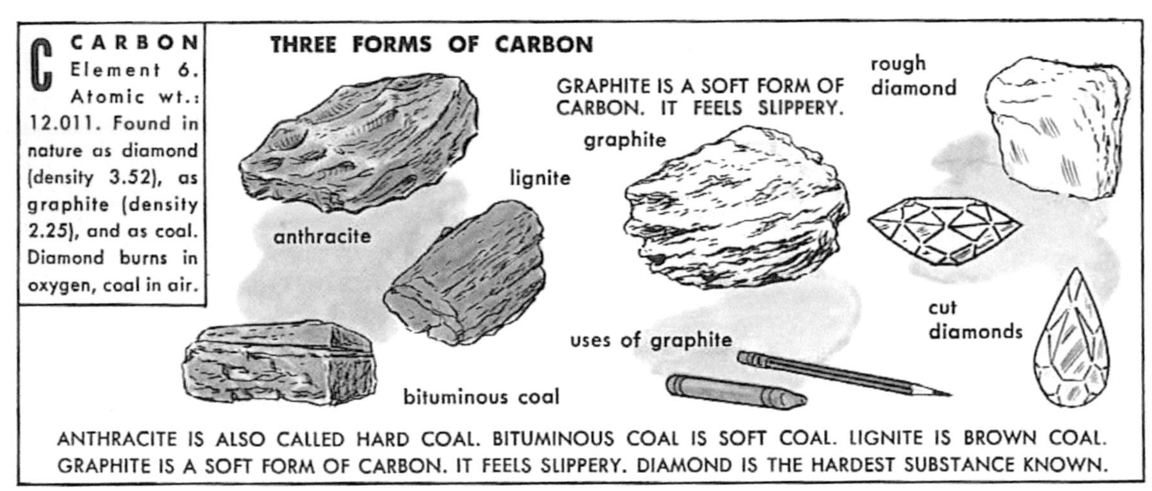

C **CARBON** Element 6. Atomic wt.: 12.011. Found in nature as diamond (density 3.52), as graphite (density 2.25), and as coal. Diamond burns in oxygen, coal in air.

ANTHRACITE IS ALSO CALLED HARD COAL. BITUMINOUS COAL IS SOFT COAL. LIGNITE IS BROWN COAL. GRAPHITE IS A SOFT FORM OF CARBON. IT FEELS SLIPPERY. DIAMOND IS THE HARDEST SUBSTANCE KNOWN.

DESTRUCTIVE DISTILLATION

IN REGULAR DISTILLATION (SEE PAGE 61), A CHEMICAL IS PURIFIED. IN DESTRUCTIVE OR DRY DISTILLATION, THE SUBSTANCE IS BROKEN INTO SEVERAL DIFFERENT CHEMICALS.

for dry distillation of wood, whittle twig into slivers, or use wooden matches without heads.

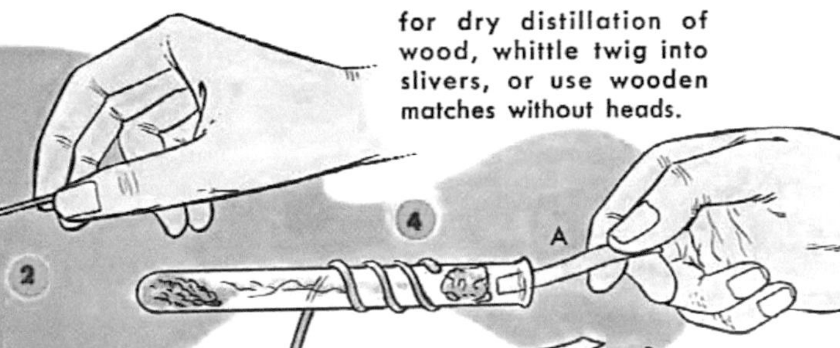

for dry distillation of coal, crush lump of bituminous coal into powder.

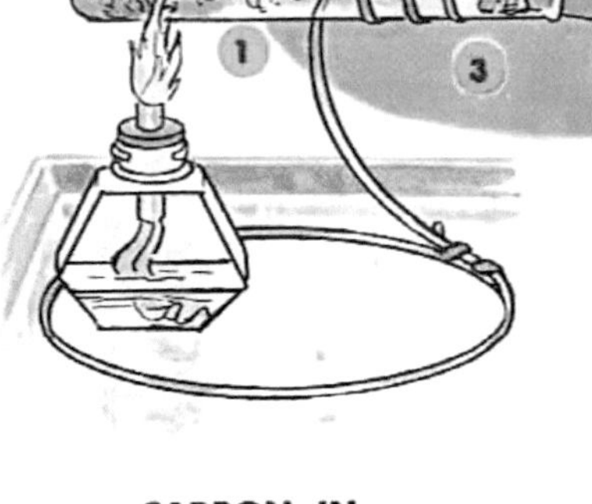

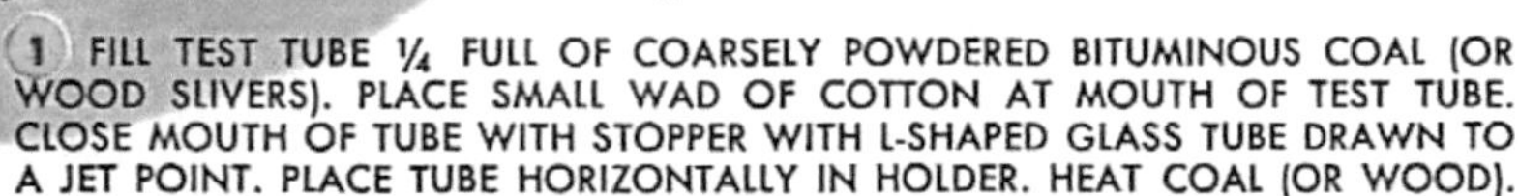

1 FILL TEST TUBE ¼ FULL OF COARSELY POWDERED BITUMINOUS COAL (OR WOOD SLIVERS). PLACE SMALL WAD OF COTTON AT MOUTH OF TEST TUBE. CLOSE MOUTH OF TUBE WITH STOPPER WITH L-SHAPED GLASS TUBE DRAWN TO A JET POINT. PLACE TUBE HORIZONTALLY IN HOLDER. HEAT COAL (OR WOOD).

2 AFTER A WHILE, DENSE FUMES DEVELOP. THEY CAN BE IGNITED AT JET.

3 COTTON WAD DISCOLORS FROM TAR CONDENSING AFTER BEING DISTILLED.

4 STOP HEATING. REMOVE STOPPER. BRING MOISTENED LITMUS PAPER TO MOUTH OF TUBE. IF YOU DISTILLED COAL, RED LITMUS TURNS BLUE FROM AMMONIA (A). IF YOU DISTILLED WOOD, BLUE LITMUS TURNS RED FROM ACETIC ACID (B). COAL HAS TURNED TO COKE, WOOD HAS BECOME CHARCOAL.

CARBON IN SUGAR

YOU CAN PROVE PRESENCE OF CARBON IN THE FOOD YOU EAT BY HEATING SMALL SAMPLES OF CHEESE, BREAD, MEAT, SUGAR. BE SURE TO DO THIS OUTDOORS TO PREVENT EXPERIMENTS FROM SMELLING UP THE WHOLE HOUSE.

HEAT 1 TEASPOON CANE SUGAR IN A CUSTARD CUP. FIRST, SUGAR MELTS. THEN IT TURNS BROWN —IT "CARAMELIZES." NEXT IT GIVES OFF THICK VAPORS THAT CAN BE IGNITED. FINALLY, A PURE FORM OF COAL REMAINS.

RUB A LUMP OF SUGAR WITH CIGARETTE ASHES (TO ACT AS CATALYST). IGNITE. DIP TEST TUBE IN LIME WATER. HOLD OVER BURNING SUGAR. FILM OF CALCIUM CARBONATE SHOWS CO_2 IN FLAME—PROVING THAT THERE IS CARBON IN SUGAR.

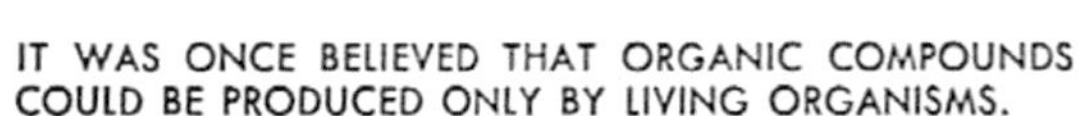

IT WAS ONCE BELIEVED THAT ORGANIC COMPOUNDS COULD BE PRODUCED ONLY BY LIVING ORGANISMS.

IN 1828, FRIEDRICH WÖHLER SUCCEEDED IN MAKING AN ORGANIC COMPOUND ARTIFICIALLY IN HIS LABORATORY.

The Chemistry of Carbon Compounds

THE CHEMISTS of about two hundred years ago divided all compounds very neatly into two groups — organic and inorganic. The organic compounds were those produced by living organisms — that is, plants and animals. The inorganic compounds were made up of dead things — rocks and minerals, water and various gases. No organic compound, these chemists insisted, could ever be produced artificially — they required the force we call "life" for their creation. And then, in 1828, a German chemist, Friedrich Wohler, completely upset this idea.

In his laboratory, Wöhler had mixed ammonium sulfate ($(NH_4)_2SO_4$) and potassium cyanate ($KCNO$), expecting to get ammonium cyanate. After evaporating, he analyzed the compound he had made. To his amazement he discovered that it was not ammonium cyanate at all, but urea — a compound produced in the kidneys of living animals, including man. The atoms of the ammonium cyanate molecule had rearranged themselves into a urea molecule.

NH_4CNO had turned into $(NH_2)_2CO$.

A few years later, another organic compound — acetic acid — was made artificially. And then the lid really blew off. More and more products of plant and animal life were put together — synthesized — in the laboratory. And as if this were not enough, chemists began producing organic compounds that were not even found in nature.

It became clear that the old meaning of organic chemistry no longer was right. And so, the definition was changed. Today, organic chemistry is defined as "the chemistry of the carbon compounds." This definition is almost, but not 100 per cent, correct. The metallic carbonates, for instance, are still considered to be inorganic compounds, and carbon dioxide and carbonic acid are regarded as being both organic and inorganic.

You may think it odd that a whole branch of chemistry should deal with the compounds of a single element. But you will not be surprised at all when you start experimenting with a few of the close to 1,000,000 carbon compounds.

HYDROCARBONS CONTAIN TWO ELEMENTS ONLY: CARBON AND HYDROGEN. HYDROCARBONS WITH FEW ATOMS TO THEIR MOLECULES ARE GASES. OTHERS WITH MANY ATOMS ARE LIQUIDS AND SOLIDS.

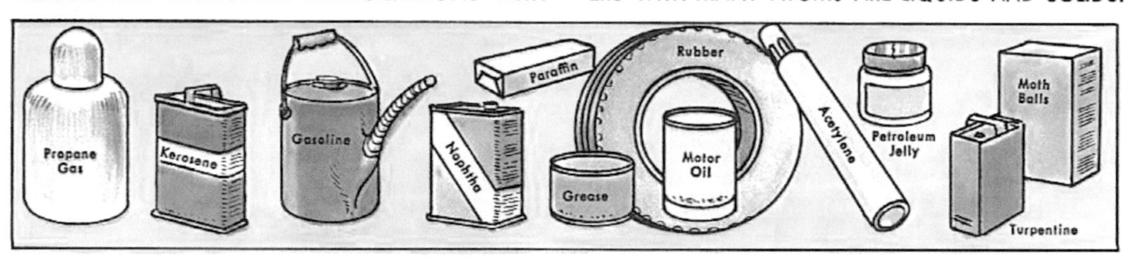

ALCOHOLS MAY BE CONSIDERED HYDROCARBONS IN WHICH A HYDROGEN ATOM IS REPLACED BY OH.

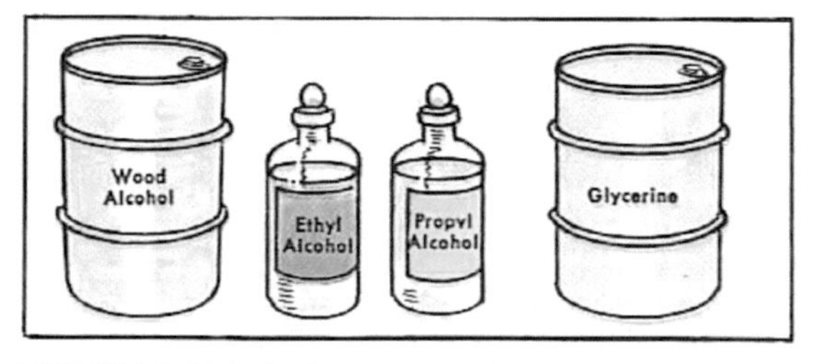

CARBOHYDRATES ARE IN MANY OF OUR MOST VALUABLE FOODSTUFFS AS STARCH AND SUGARS.

ESTERS IN ORGANIC CHEMISTRY CAN BE COMPARED TO SALTS IN INORGANIC CHEMISTRY. FATS AND OILS ARE THE MOST IMPORTANT ESTERS. THESE ARE THE "SALTS" OF GLYCERINE AND FATTY ACIDS.

CARBOXYLIC ACIDS ARE NAMED FOR THE CARBOXYL GROUP—COOH—FOUND IN THEIR FORMULAS.

PROTEINS ARE COMPLEX COMPOUNDS THAT CONTAIN CARBON, HYDROGEN, OXYGEN, NITROGEN.

OTHER CARBON COMPOUNDS—IN ADDITION TO THE MAIN GROUPS ILLUSTRATED ABOVE, THERE ARE NUMEROUS OTHER KINDS OF CARBON COMPOUNDS. MANY HAVE VERY COMPLICATED FORMULAS.

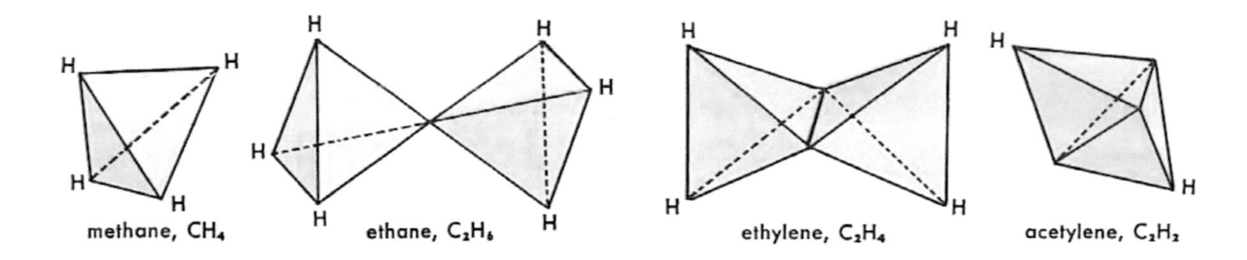

The Formulas of Carbon Compounds

How is it possible for carbon to make so many different compounds of such tremendous variety? That was one of the great questions facing chemists during the last century.

It was easy enough to explain carbon dioxide. Carbon has a valence of 4, oxygen of 2 — the formula had to be CO_2. It was also easy to explain the molecule of the simple hydrocarbon methane (CH_4). But how explain compounds consisting of two atoms of carbon and six of hydrogen (C_2H_6, ethane), or two atoms of carbon and four of hydrogen (C_2H_4, ethylene), or two of carbon and only two of hydrogen (C_2H_2, acetylene)?

A German chemist and professor, Friedrich August Kekulé, came up with the solution. The answer was quite simple:

While the atoms of most elements "hook on" to the atoms of other elements according to their valences, the atoms of carbon "hook on" to each other as well. To understand this, write out carbon atoms with four lines to indicate the valence bonds, but arrange the lines in these three different ways:

$$\equiv C- \quad -C\equiv \quad =C=$$

Then hook them together, two by two, in these three different ways:

$$\equiv C-C\equiv \quad =C=C= \quad -C\equiv C-$$

Now add a hydrogen atom to each of the free bonds — and there you have the formulas for the three hydro-carbons — ethane (C_2H_6), ethylene (C_2H_4), and acetylene (C_2H_2):

$$H_3C-CH_3 \qquad H_2C=CH_2 \qquad H-C\equiv C-H$$

So far so good. But there were still many carbon compound formulas that would not line up in this kind of arrangement. C_6H_6, for instance — benzene, an important hydrocarbon obtained by distillation of coal.

Again, it was Kekulé who offered the explanation. This time it came to him in a dream. He had been

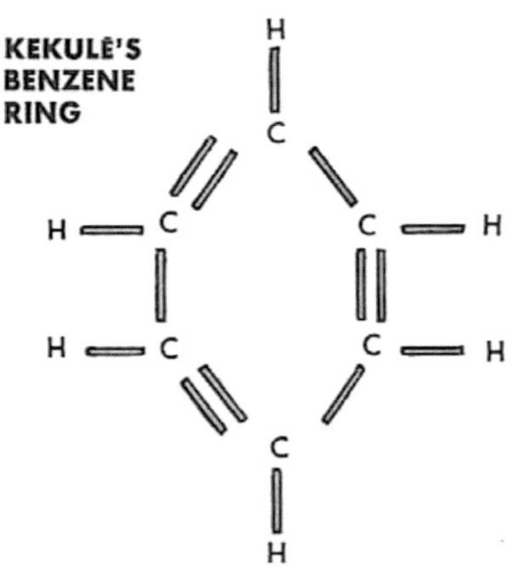

AUGUST KEKULE HIT UPON THE STRUCTURE OF THE BENZENE MOLECULE IN A DREAM. A SNAKE SEEMED TO WHIRL IN A RING BEFORE HIS EYES. BY ARRANGING THE SIX CARBON ATOMS IN A RING, THE PROBLEM WAS SOLVED.

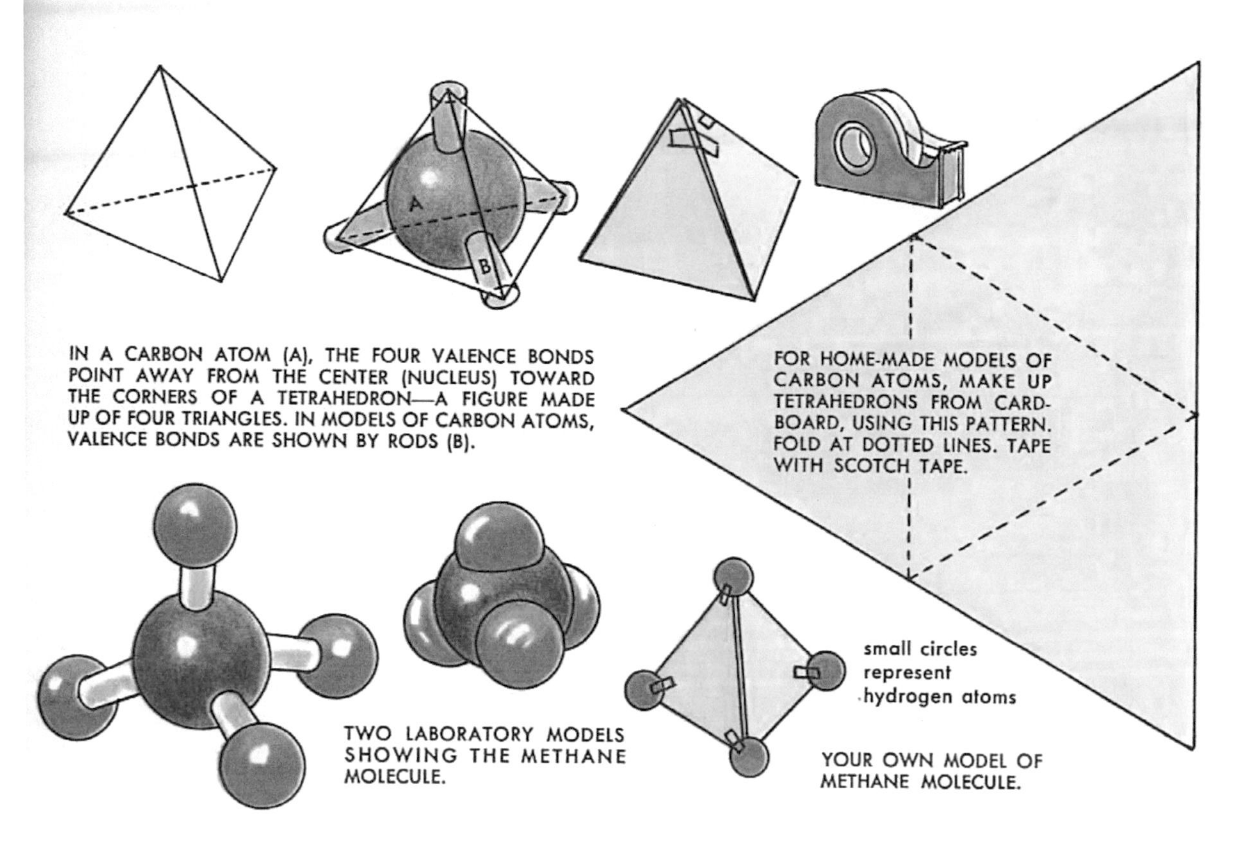

IN A CARBON ATOM (A), THE FOUR VALENCE BONDS POINT AWAY FROM THE CENTER (NUCLEUS) TOWARD THE CORNERS OF A TETRAHEDRON—A FIGURE MADE UP OF FOUR TRIANGLES. IN MODELS OF CARBON ATOMS, VALENCE BONDS ARE SHOWN BY RODS (B).

FOR HOME-MADE MODELS OF CARBON ATOMS, MAKE UP TETRAHEDRONS FROM CARDBOARD, USING THIS PATTERN. FOLD AT DOTTED LINES. TAPE WITH SCOTCH TAPE.

TWO LABORATORY MODELS SHOWING THE METHANE MOLECULE.

YOUR OWN MODEL OF METHANE MOLECULE.

working all day with long lines of organic formulas. In the evening he dozed before the fire. In his dream, the lines of formulas turned into snakes, twisting and twining — until suddenly one of the snakes grasped its own tail and whirled around in a ring. This dream gave Kekule the clue: the carbon atoms in benzene hang together in a ring, each atom using three of its bonds to hang on to the atoms next to it, with one bond free to hook onto a hydrogen atom.

Starting from these very simple formulas, modern scientists can figure out the most complicated chemical formulas.

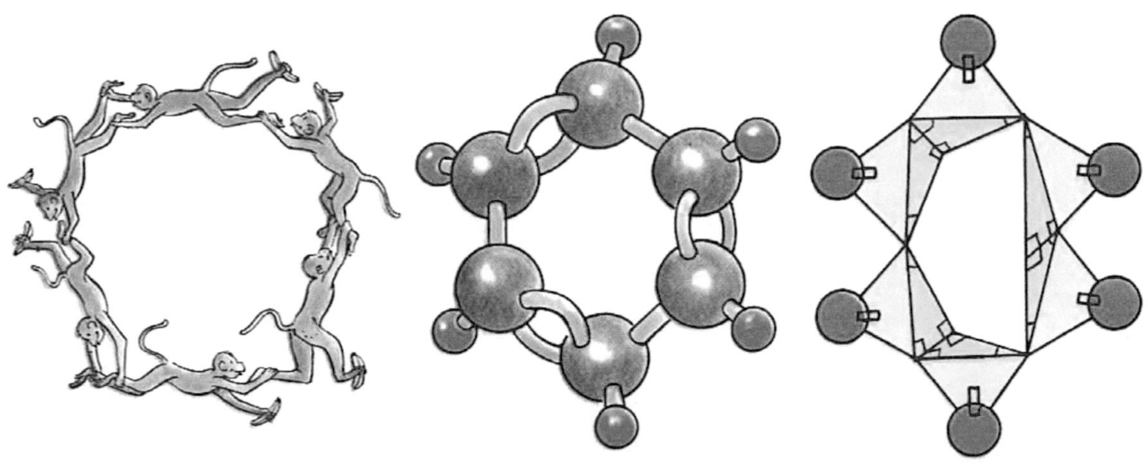

YOU CAN THINK OF THE BENZENE RING AS SIX MONKEYS HANGING ON TO EACH OTHER WITH ONE OR TWO HANDS, HOLDING BANANAS IN THEIR FREE HANDS.

THIS IS THE WAY THE BENZENE MOLECULE LOOKS WHEN IT IS CONSTRUCTED FROM PARTS USED TO MAKE UP LABORATORY MODELS FOR DEMONSTRATION.

THIS IS HOW THE BENZENE MOLECULE WILL LOOK WHEN YOU P' TOGETHER FROM HOME-MADE BON ATOMS. YOU CAN DO WITH SCOTCH TAPE.

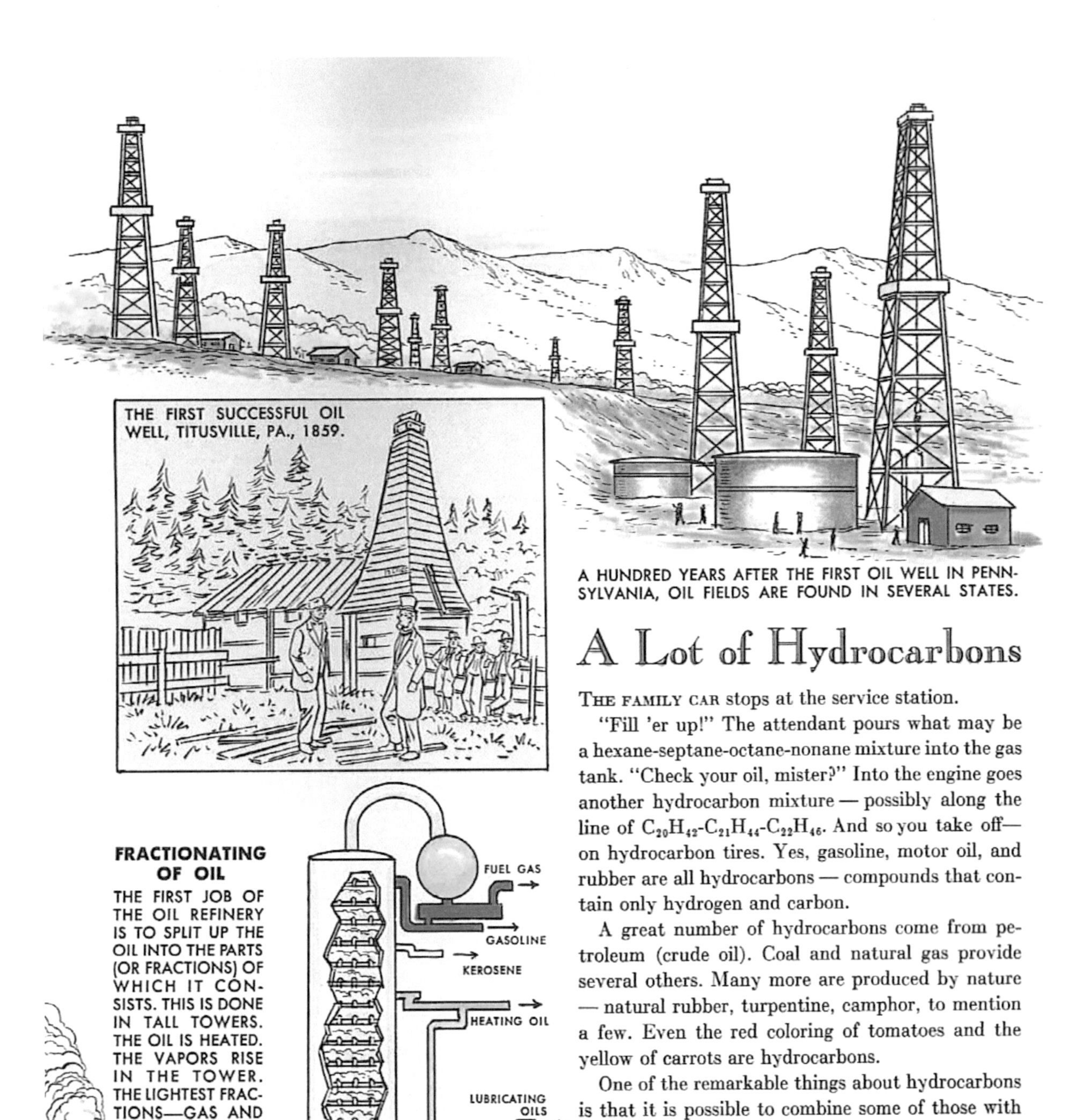

THE FIRST SUCCESSFUL OIL WELL, TITUSVILLE, PA., 1859.

A HUNDRED YEARS AFTER THE FIRST OIL WELL IN PENNSYLVANIA, OIL FIELDS ARE FOUND IN SEVERAL STATES.

FRACTIONATING OF OIL

THE FIRST JOB OF THE OIL REFINERY IS TO SPLIT UP THE OIL INTO THE PARTS (OR FRACTIONS) OF WHICH IT CONSISTS. THIS IS DONE IN TALL TOWERS. THE OIL IS HEATED. THE VAPORS RISE IN THE TOWER. THE LIGHTEST FRACTIONS—GAS AND GASOLINE—GO TO THE TOP, FOLLOWED BY KEROSENE, FUEL OIL, LUBRICATING OILS, WAX, ASPHALT.

A Lot of Hydrocarbons

THE FAMILY CAR stops at the service station.

"Fill 'er up!" The attendant pours what may be a hexane-septane-octane-nonane mixture into the gas tank. "Check your oil, mister?" Into the engine goes another hydrocarbon mixture — possibly along the line of $C_{20}H_{42}$-$C_{21}H_{44}$-$C_{22}H_{46}$. And so you take off—on hydrocarbon tires. Yes, gasoline, motor oil, and rubber are all hydrocarbons — compounds that contain only hydrogen and carbon.

A great number of hydrocarbons come from petroleum (crude oil). Coal and natural gas provide several others. Many more are produced by nature — natural rubber, turpentine, camphor, to mention a few. Even the red coloring of tomatoes and the yellow of carrots are hydrocarbons.

One of the remarkable things about hydrocarbons is that it is possible to combine some of those with small molecules into others with larger ones (as in making synthetic rubber), as well as to "crack" those with large molecules into others with smaller ones (as when a heavy oil is "cracked" into gasoline). But that is only the beginning. By replacing one or more hydrogen atoms with hydroxyl groups (OH) or carboxyl groups (COOH) or chlorine atoms (Cl), for instance, it is possible to build up more complicated compounds — which can then be built up further and further. And that is exactly what chemists are doing today — giving us medicines and dyes, plastics and explosives, and countless other things.

1 HAMMER LUMPS OF BITUMINOUS COAL INTO A COARSE POWDER. FILL FUNNEL WITH IT. BRING FUNNEL INTO LARGE JAR.

2 TURN JAR UPSIDE DOWN. FILL JAR WITH WATER. PLACE A WATER-FILLED TEST TUBE OVER FUNNEL. IN A FEW DAYS, TUBE IS FILLED WITH METHANE.

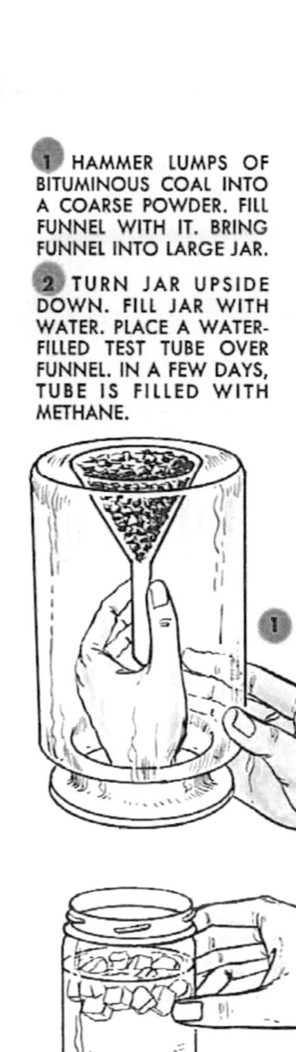

METHANE—CH_4

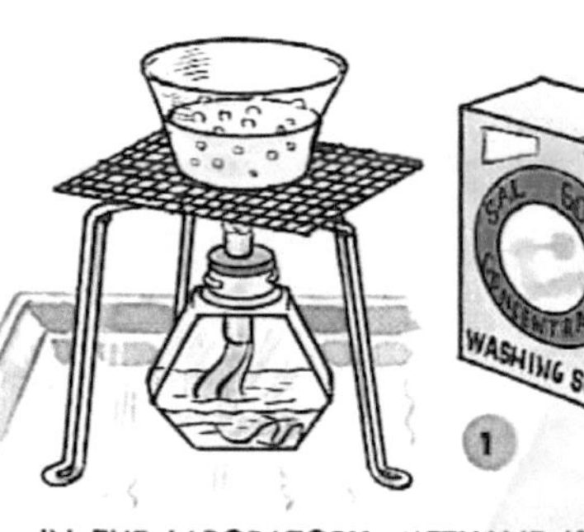

IN THE LABORATORY, METHANE IS MADE BY HEATING WATER-FREE SODIUM ACETATE WITH "SODA LIME."

1 TO MAKE SODIUM ACETATE, ADD WASHING SODA TO ½ CUSTARD CUP VINEGAR UNTIL NO MORE CO_2 IS GIVEN OFF. EVAPORATE MIXTURE AT LOW HEAT TO GET WHITE POWDER OF SODIUM ACETATE.

2 MIX 5 g SODIUM ACETATE (CH_3COONa), 5 g SODIUM HYDROXIDE, AND 5 g CALCIUM OXIDE. DROP INTO TEST TUBE. SET UP APPARATUS FOR COLLECTING GAS AS SHOWN BELOW. HEAT TO MAKE METHANE:

$$CH_3COONa + NaOH \rightarrow CH_4 + Na_2CO_3$$

NAPHTHALENE—$C_{10}H_8$

NAPHTHALENE IS USED IN MAKING MOTH BALLS.

NAPHTHALENE CAN BE PURIFIED BY SUBLIMATION. TO DEMONSTRATE THIS, CRUSH A COUPLE OF MOTH BALLS. HEAT THEM IN A CUSTARD CUP. FIRST THEY MELT, THEN GIVE OFF VAPOR. PUT JAR FILLED WITH ICE WATER OVER CUP. NAPHTHALENE SETTLES ON BOTTOM IN LEAFY CRYSTALS.

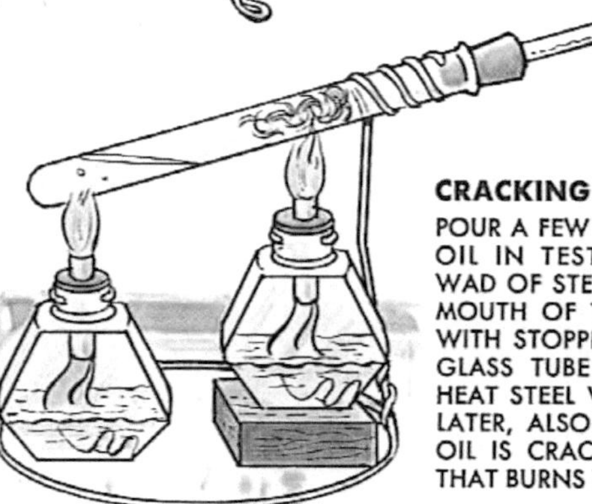

CRACKING OIL

POUR A FEW ml HOUSEHOLD OIL IN TEST TUBE. PLACE WAD OF STEEL WOOL NEAR MOUTH OF TUBE. CLOSE IT WITH STOPPER THAT HAS A GLASS TUBE WITH JET TIP. HEAT STEEL WOOL. A LITTLE LATER, ALSO HEAT THE OIL. OIL IS CRACKED INTO GAS THAT BURNS WHEN IGNITED.

TURPENTINE—$C_{10}H_{16}$

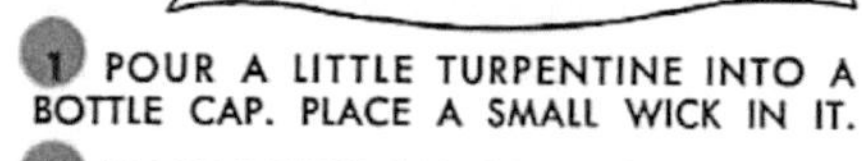

1 POUR A LITTLE TURPENTINE INTO A BOTTLE CAP. PLACE A SMALL WICK IN IT.

2 PLACE BOTTLE CAP ON PIECE OF PAPER. IGNITE TURPENTINE. IT BURNS INCOMPLETELY, GIVING OFF A BLACK SMOKE OF CARBON WHICH YOU CAN COLLECT IN A JAR.

MUCH OF THE SUGAR WE USE IS MADE BY EVAPORATING THE JUICE OF SUGAR BEETS AND SUGAR CANE. MAPLE SUGAR IS BOILED-DOWN SAP OF SUGAR MAPLE TREES.

Carbohydrates—Sweet and Bland

USUALLY, when we talk about "hydrates" we mean chemicals that contain water. But when we talk about carbohydrates we mean organic compounds of carbon, hydrogen, and oxygen in which the proportion between hydrogen and oxygen is the same as in water (H_2O) — that is, twice as much hydrogen as oxygen. And so we find carbohydrates that have 22 atoms of hydrogen and 11 atoms of oxygen to 12 atoms of carbon ($C_{12}H_{22}O_{11}$), or 12 hydrogen and 6 oxygen to 6 carbon ($C_6H_{12}O_6$), or 10 and 5 to 6 carbon atoms ($C_6H_{10}O_5$).

THE SWEETNESS OF FRUITS AND BERRIES COMES FROM A MIXTURE OF TWO KINDS OF SUGAR CALLED FRUCTOSE AND GLUCOSE. THESE SUGARS ARE MADE IN THE GREEN LEAVES OF THE PLANT AND SENT INTO THE FRUITS FOR STORAGE.

Carbohydrates are produced by plants by a remarkable process called photosynthesis — "putting things together with the help of light." When green leaves are exposed to sunlight, the chlorophyll in them combines the hydrogen from water with carbon dioxide from the air, while setting oxygen free — along this line:

$$6H_2O + 6CO_2 + \text{sunlight} \rightarrow C_6H_{12}O_6 + 6O_2 \uparrow$$

Carbohydrates are of tremendous importance to all of us. They make up a large part of our food supply in the form of sugars and starches. Another carbohydrate called cellulose helps to clothe us (cotton, linen) and shelter us (wood).

SUGARS — Most of our sugar comes from sugar beets or sugar cane. The juice is pressed out, cleared, filtered, and evaporated. The result is pure, white crystals of a sugar with the chemical name sucrose ($C_{12}H_{22}O_{11}$).

Another sugar called glucose ($C_6H_{12}O_6$) is found in ripe fruits, often in the company of still another sugar of the same formula called fructose ($C_6H_{12}O_6$). These two sugars can be made in the laboratory by treating the more complicated sucrose with an acid. The sucrose picks up water and splits into glucose and fructose by a process known as inversion:

$$\underset{\text{(sucrose)}}{C_{12}H_{22}O_{11}} + H_2O \rightarrow \underset{\text{(glucose)}}{C_6H_{12}O_6} + \underset{\text{(fructose)}}{C_6H_{12}O_6}$$

(CONTINUED ON PAGE 86)

TEST FOR GLUCOSE SUGAR

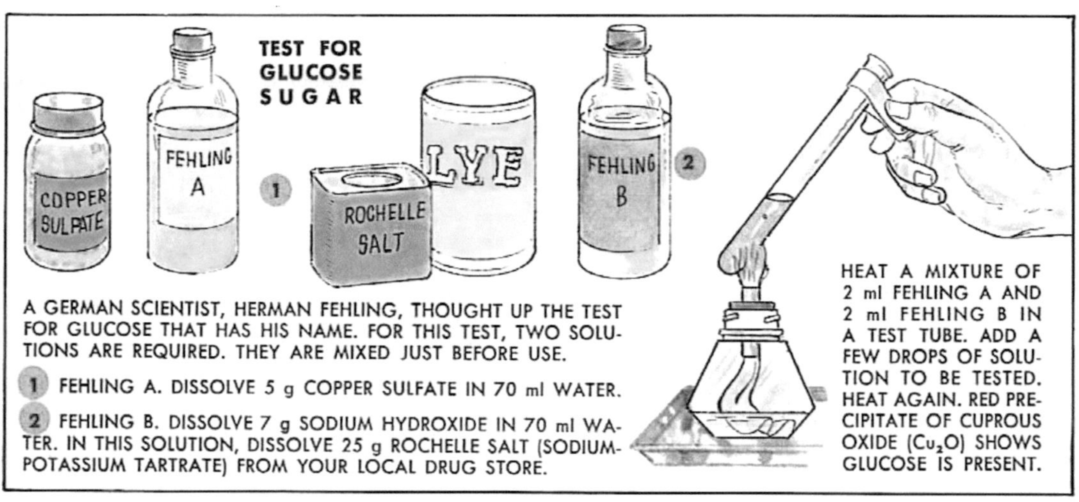

A GERMAN SCIENTIST, HERMAN FEHLING, THOUGHT UP THE TEST FOR GLUCOSE THAT HAS HIS NAME. FOR THIS TEST, TWO SOLUTIONS ARE REQUIRED. THEY ARE MIXED JUST BEFORE USE.

1 FEHLING A. DISSOLVE 5 g COPPER SULFATE IN 70 ml WATER.

2 FEHLING B. DISSOLVE 7 g SODIUM HYDROXIDE IN 70 ml WATER. IN THIS SOLUTION, DISSOLVE 25 g ROCHELLE SALT (SODIUM-POTASSIUM TARTRATE) FROM YOUR LOCAL DRUG STORE.

HEAT A MIXTURE OF 2 ml FEHLING A AND 2 ml FEHLING B IN A TEST TUBE. ADD A FEW DROPS OF SOLUTION TO BE TESTED. HEAT AGAIN. RED PRECIPITATE OF CUPROUS OXIDE (Cu_2O) SHOWS GLUCOSE IS PRESENT.

TEST CANE SUGAR WITH FEHLING. YOU DO NOT GET RED PRECIPITATE. CANE SUGAR IS NOT GLUCOSE BUT ANOTHER SUGAR CALLED SUCROSE.

USE FEHLING TEST TO FIND OUT IF DIFFERENT SWEET-TASTING FOODS CONTAIN GLUCOSE SUGAR: CORN SYRUP, MAPLE SYRUP, MOLASSES, HONEY. ALSO TRY JUICES OF VARIOUS FRUITS: PRUNES, ORANGES, LEMONS, BERRIES. SEVERAL CONTAIN GLUCOSE AND GIVE RED PRECIPITATE. SUGAR IN MILK (LACTOSE) GIVES Cu_2O PRECIPITATE.

SUCROSE TO GLUCOSE

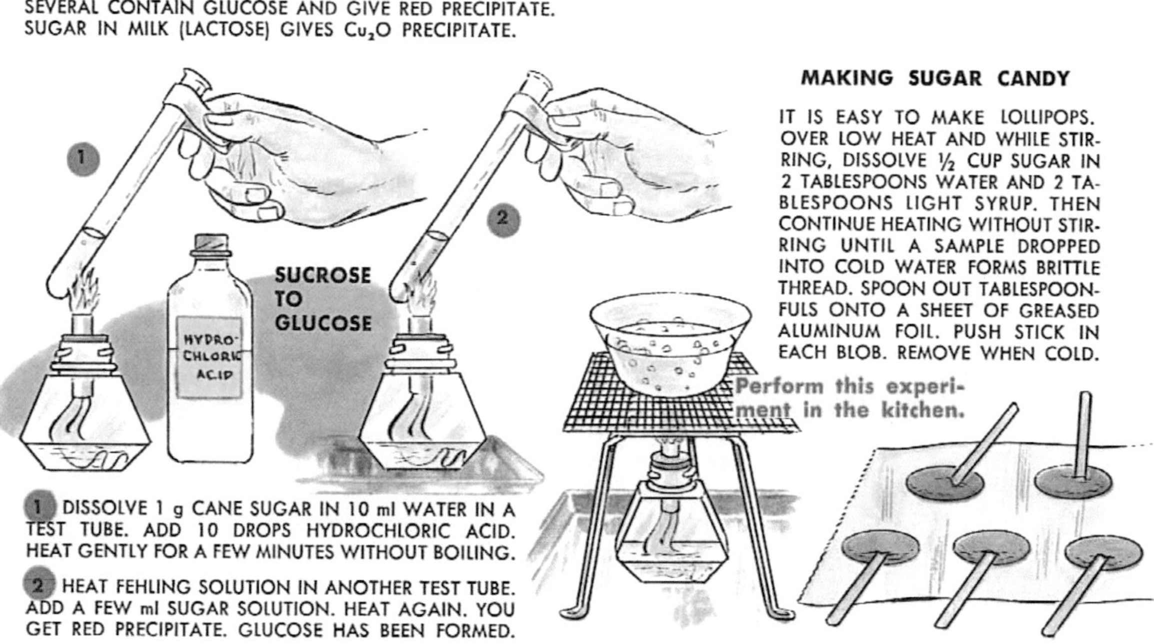

1 DISSOLVE 1 g CANE SUGAR IN 10 ml WATER IN A TEST TUBE. ADD 10 DROPS HYDROCHLORIC ACID. HEAT GENTLY FOR A FEW MINUTES WITHOUT BOILING.

2 HEAT FEHLING SOLUTION IN ANOTHER TEST TUBE. ADD A FEW ml SUGAR SOLUTION. HEAT AGAIN. YOU GET RED PRECIPITATE. GLUCOSE HAS BEEN FORMED.

MAKING SUGAR CANDY

IT IS EASY TO MAKE LOLLIPOPS. OVER LOW HEAT AND WHILE STIRRING, DISSOLVE ½ CUP SUGAR IN 2 TABLESPOONS WATER AND 2 TABLESPOONS LIGHT SYRUP. THEN CONTINUE HEATING WITHOUT STIRRING UNTIL A SAMPLE DROPPED INTO COLD WATER FORMS BRITTLE THREAD. SPOON OUT TABLESPOONFULS ONTO A SHEET OF GREASED ALUMINUM FOIL. PUSH STICK IN EACH BLOB. REMOVE WHEN COLD.

Perform this experiment in the kitchen.

Carbohydrates—Continued

STARCHES — Starch is distributed in most plant parts. It is a carbohydrate with very large molecules. Take a look at its formula: $(C_6H_{10}O_5)_x$. At first glance it looks quite simple. But note that little x — it stands for "any number of times." A single molecule of starch may weigh 6,000 times as much as a single molecule of glucose.

You can break this polysaccharide ("many-sugar") into the monosaccharide ("single-sugar") glucose by treating it with an acid.

CELLULOSE is the building material of the plant world. It makes up the cell walls of leaves and stalks, wood and fibers. Cotton is 95 per cent cellulose. The paper on which this book is printed is specially treated cellulose. So is the cellophane around your candy and the rayon that goes into ladies' dresses. For more about cellulose in natural fibers and rayon, see pages 102-103.

A GROWING PLANT IS THE MOST ASTONISHING CHEMICAL FACTORY ON EARTH. THE GREEN SUBSTANCE IN LEAVES—CALLED CHLOROPHYLL—WITH THE HELP OF SUNLIGHT IS ABLE TO COMBINE WATER (TAKEN IN BY THE ROOTS) WITH CARBON DIOXIDE FROM THE AIR (TAKEN IN THROUGH THE LEAVES) TO FORM SUGAR FIRST AND THEN STARCH.

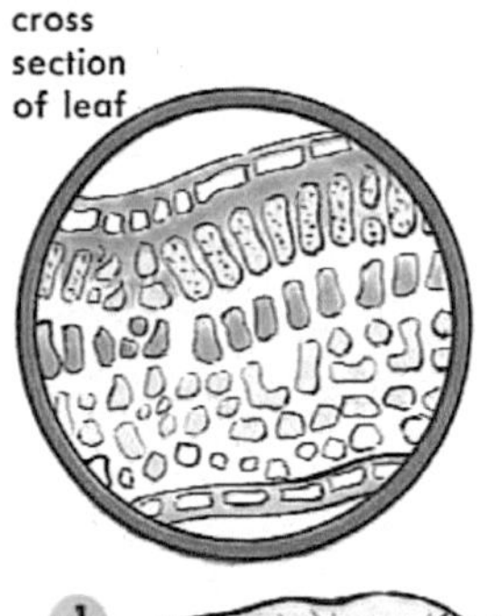

cross section of leaf

EXPERIMENTS WITH PHOTOSYNTHESIS

1 POT UP A NASTURTIUM OR GERANIUM PLANT AND PLACE IT IN THE DARK FOR A COUPLE OF DAYS. THEN FASTEN STRIPS OF BLACK PAPER ACROSS BOTH SIDES OF ONE OR MORE LEAVES. NOW EXPOSE THE GROWING PLANT TO THE SUNLIGHT FOR TWO HOURS.

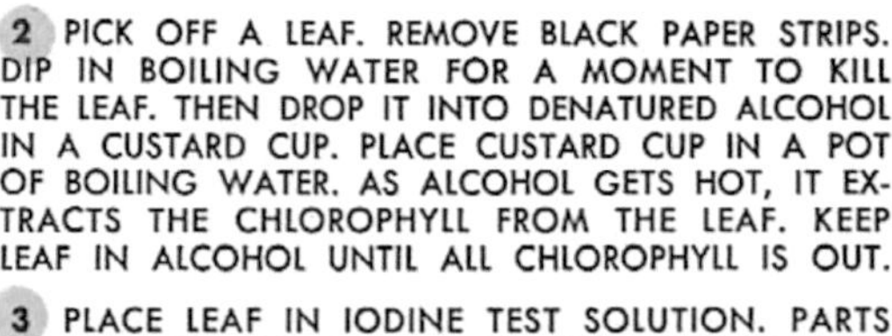

2 PICK OFF A LEAF. REMOVE BLACK PAPER STRIPS. DIP IN BOILING WATER FOR A MOMENT TO KILL THE LEAF. THEN DROP IT INTO DENATURED ALCOHOL IN A CUSTARD CUP. PLACE CUSTARD CUP IN A POT OF BOILING WATER. AS ALCOHOL GETS HOT, IT EXTRACTS THE CHLOROPHYLL FROM THE LEAF. KEEP LEAF IN ALCOHOL UNTIL ALL CHLOROPHYLL IS OUT.

3 PLACE LEAF IN IODINE TEST SOLUTION. PARTS EXPOSED TO SUN TURN BLUE. THIS PROVES PRESENCE OF STARCH. UNEXPOSED PARTS BECOME BROWN.

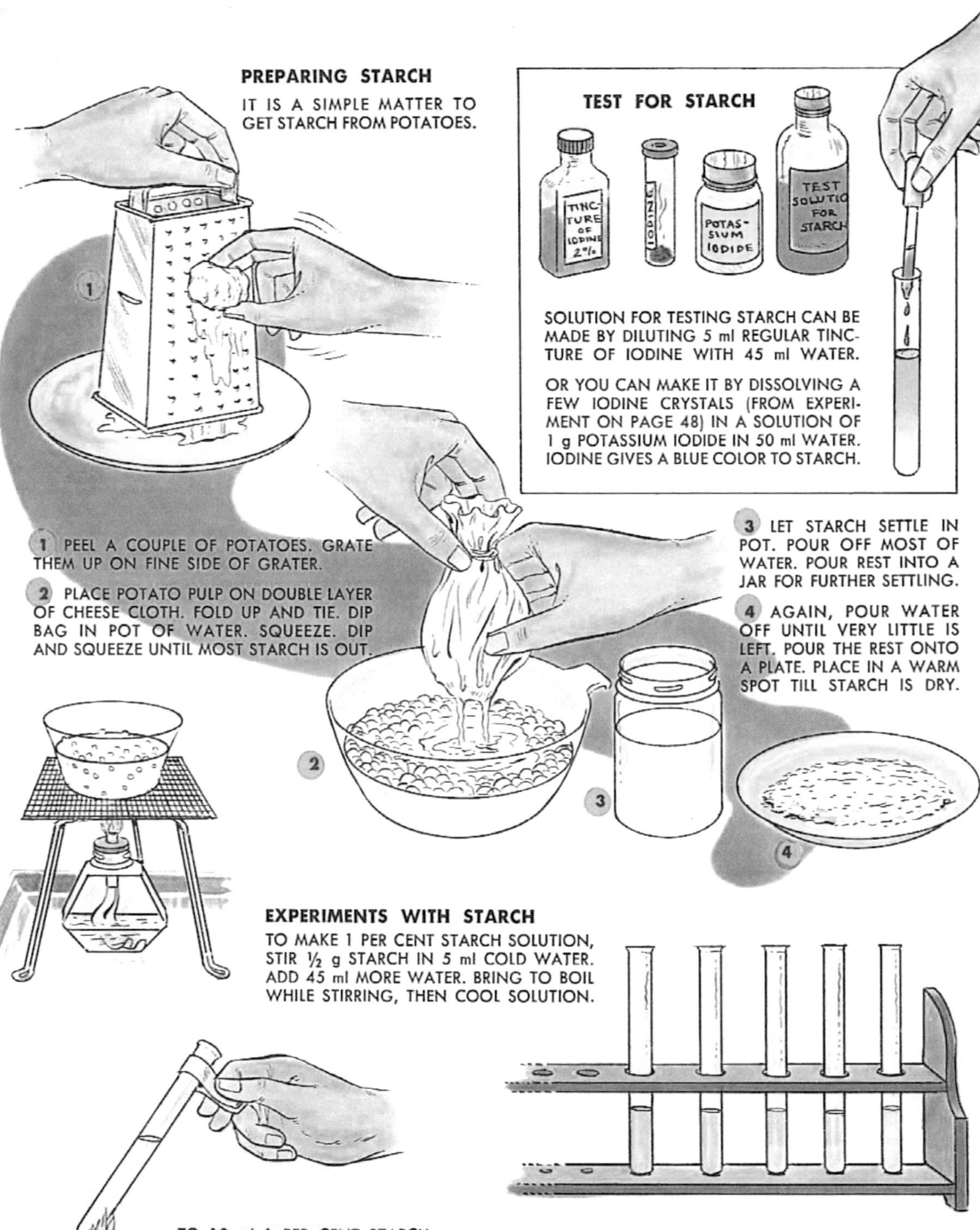

PREPARING STARCH

IT IS A SIMPLE MATTER TO GET STARCH FROM POTATOES.

TEST FOR STARCH

SOLUTION FOR TESTING STARCH CAN BE MADE BY DILUTING 5 ml REGULAR TINCTURE OF IODINE WITH 45 ml WATER.

OR YOU CAN MAKE IT BY DISSOLVING A FEW IODINE CRYSTALS (FROM EXPERIMENT ON PAGE 48) IN A SOLUTION OF 1 g POTASSIUM IODIDE IN 50 ml WATER. IODINE GIVES A BLUE COLOR TO STARCH.

1 PEEL A COUPLE OF POTATOES. GRATE THEM UP ON FINE SIDE OF GRATER.

2 PLACE POTATO PULP ON DOUBLE LAYER OF CHEESE CLOTH. FOLD UP AND TIE. DIP BAG IN POT OF WATER. SQUEEZE. DIP AND SQUEEZE UNTIL MOST STARCH IS OUT.

3 LET STARCH SETTLE IN POT. POUR OFF MOST OF WATER. POUR REST INTO A JAR FOR FURTHER SETTLING.

4 AGAIN, POUR WATER OFF UNTIL VERY LITTLE IS LEFT. POUR THE REST ONTO A PLATE. PLACE IN A WARM SPOT TILL STARCH IS DRY.

EXPERIMENTS WITH STARCH

TO MAKE 1 PER CENT STARCH SOLUTION, STIR ½ g STARCH IN 5 ml COLD WATER. ADD 45 ml MORE WATER. BRING TO BOIL WHILE STIRRING, THEN COOL SOLUTION.

TO 10 ml 1 PER CENT STARCH SOLUTION ADD 10 DROPS HYDROCHLORIC ACID. BOIL FOR 2 MINUTES. TEST THE RESULT WITH FEHLING SOLUTION. YOU GET RED PRECIPITATE THAT SHOWS PRESENCE OF GLUCOSE. UNTREATED STARCH SOLUTION DOES NOT REACT WITH THE FEHLING SOLUTION.

LINE UP FIVE TEST TUBES, EACH CONTAINING 5 ml WATER AND 1 DROP IODINE TEST SOLUTION. IN ANOTHER TEST TUBE, ADD 2 DROPS OF SALIVA (SPITTLE) TO 5 ml STARCH SOLUTION. PLACE THIS IN GLASS OF WARM (NOT HOT) WATER. WITH 2-MINUTE INTERVALS, DROP 3 DROPS SALIVA-STARCH MIXTURE INTO A TEST TUBE WITH IODINE SOLUTION. SHAKE. COLOR GETS LESS AND LESS BLUE. SALIVA DIGESTS THE STARCH AND TURNS IT INTO A SUGAR, MALTOSE.

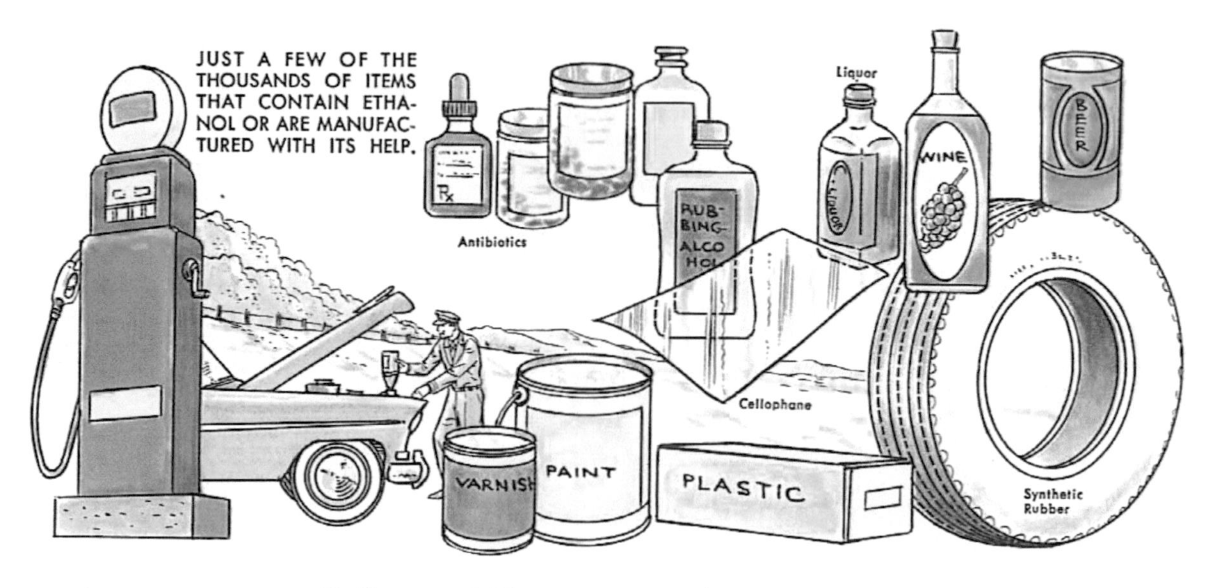

Many Kinds of Alcohols

TO MOST PEOPLE, alcohol is the strong stuff in beer, wine, and hard liquor. But to a chemist, this is just one of many alcohols.

Alcohols may be considered hydrocarbons in which one or more hydrogen (H) atoms are replaced by hydroxyl (OH) groups. Their names are made up from the names of the hydrocarbons to which they are related by giving these an "-ol" ending. In this way, CH_4, methane, becomes CH_3OH, methanol (also called methyl alcohol); C_2H_6, ethane, becomes C_2H_5OH, ethanol (also known as ethyl or grain alcohol); and so on. Methanol (CH_3OH) was originally called wood alcohol because it was made by the destructive distillation of wood. It is very poisonous and is therefore used to "denature" ethanol, making this unfit for drinking.

Ethanol (C_2H_5OH) is produced today, to a great extent, in the same way in which it was made thousands of years ago, by a process called fermentation. In this, the tiny plant cells of yeast are made to grow in the solution of a simple sugar, such as glucose ($C_6H_{12}O_6$). In growing, the yeast cells give off a substance called zymase. This acts as a catalyst and turns the glucose into ethanol and carbon dioxide:

$$C_6H_{12}O_6 \rightarrow 2C_2H_5OH + 2CO_2 \uparrow$$

The ethanol is finally separated from the watery liquid by distillation.

Glycerol ($C_3H_5(OH)_3$) is still another alcohol which you probably know better under the name of glycerin. Glycerol may be considered a product of propane (C_3H_8) in which not one but three H atoms have been replaced by OH.

THE "FAMILY TREE" OF ETHANOL—WITH SOME OF ITS CHILDREN, GRANDCHILDREN, AND GREAT-GRANDCHILDREN.

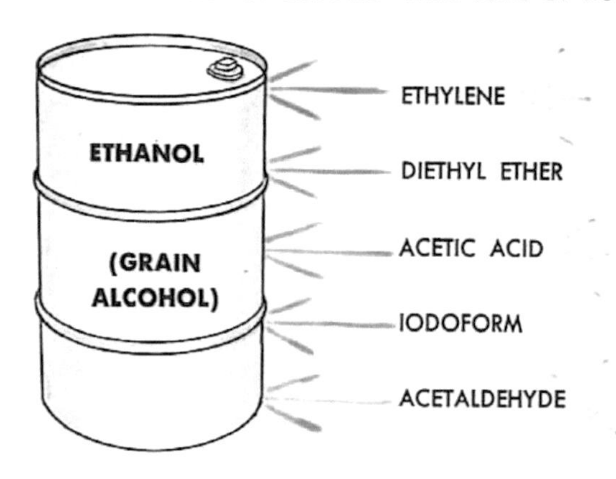

STYRENE — SYNTHETIC RUBBER, PLASTICS

GLYCOL — OXALIC ACID, EXPLOSIVES

ACETONE — BUTYL ALCOHOL, CHLOROFORM

CELLULOSE ACETATE — PHOTOGRAPHIC FILM, TEXTILE FIBERS

ACETALDEHYDE CYANOHYDRIN — PROPIONIC ACID

ACETIC ANHYDRIDE — VINYL ALCOHOL, ACETANILIDE

METHYL ALCOHOL—METHANOL

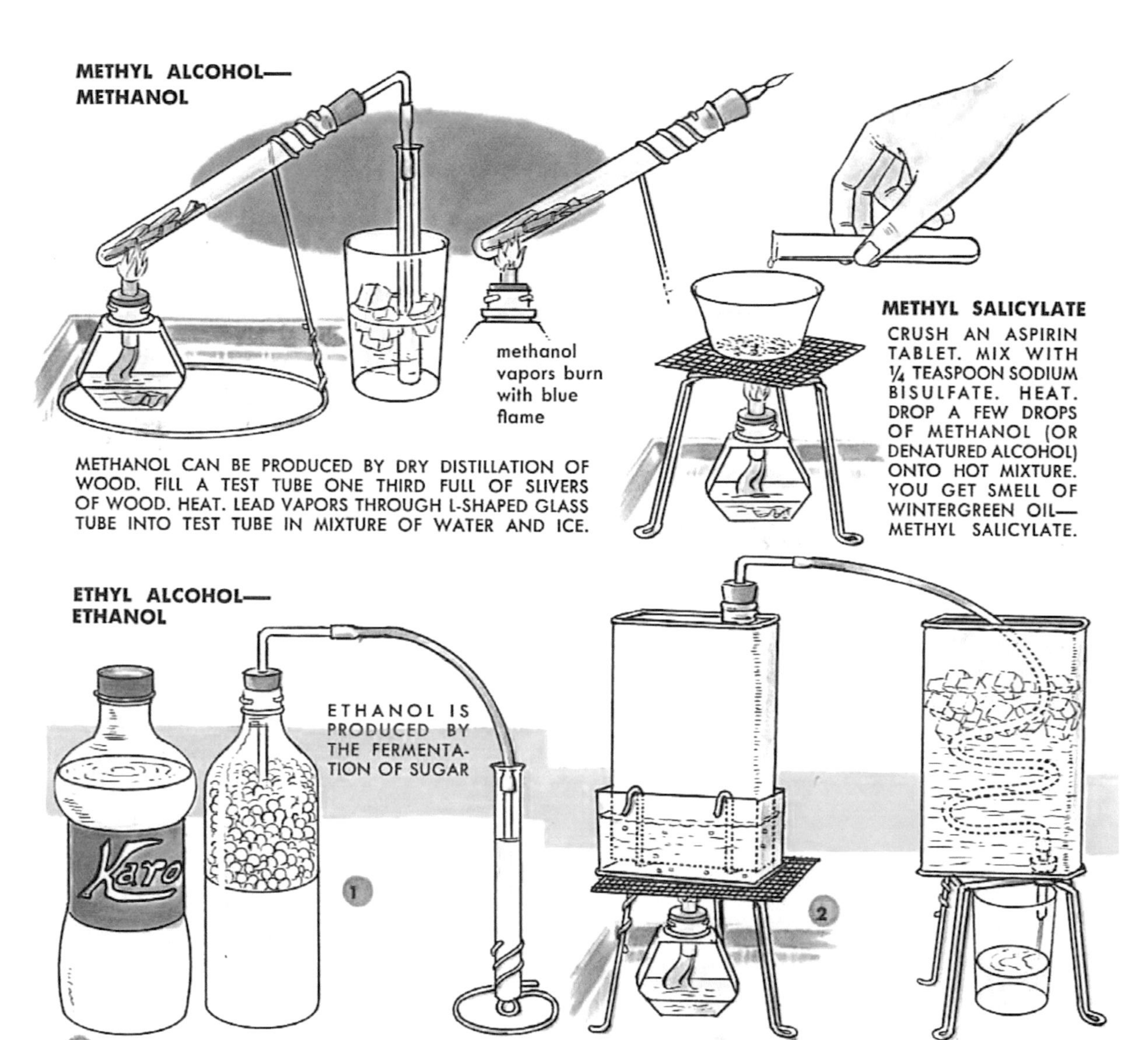

METHANOL CAN BE PRODUCED BY DRY DISTILLATION OF WOOD. FILL A TEST TUBE ONE THIRD FULL OF SLIVERS OF WOOD. HEAT. LEAD VAPORS THROUGH L-SHAPED GLASS TUBE INTO TEST TUBE IN MIXTURE OF WATER AND ICE.

METHYL SALICYLATE

CRUSH AN ASPIRIN TABLET. MIX WITH ¼ TEASPOON SODIUM BISULFATE. HEAT. DROP A FEW DROPS OF METHANOL (OR DENATURED ALCOHOL) ONTO HOT MIXTURE. YOU GET SMELL OF WINTERGREEN OIL—METHYL SALICYLATE.

ETHYL ALCOHOL—ETHANOL

1 IN A PINT BOTTLE MIX ¼ CUP CORN SYRUP WITH 1 CUP WARM WATER. ADD ½ PACKAGE YEAST THAT HAS BEEN SOFTENED IN LUKEWARM WATER. PLACE BOTTLE IN A WARM SPOT. SHORTLY THE LIQUID BEGINS TO BUBBLE. LEAD THE GAS INTO LIME WATER. GAS IS CO_2. IN A FEW DAYS, GAS DEVELOPMENT SLOWS DOWN.

2 FILTER HALF OF THE FERMENTED LIQUID INTO A 1-PINT SCREW-TOP CAN. SET UP APPARATUS FOR DISTILLATION AS DESCRIBED ON PAGE 61 WITH THE EXCEPTION THAT HEATING IS DONE ON A WATER BATH MADE FROM HALF A QUART CAN WITH WATER. DISTILL OFF A FEW ml ETHANOL AT LOWEST POSSIBLE HEAT.

IODOFORM FROM ETHANOL

TO A SOLUTION OF 1 g POTASSIUM IODIDE IN 5 ml WATER ADD IODINE CRYSTALS TO GET DARK BROWN COLOR. ADD 5 ml ETHANOL. ADD 10% NaOH SOLUTION UNTIL COLOR DISAPPEARS. HEAT GENTLY TWO MINUTES. LET COOL. THE YELLOW PRECIPITATE IS IODOFORM—CHI_3.

ETHYL ACETATE FROM ETHANOL

IN A TEST TUBE, MIX 3 ml ETHANOL WITH 2 g SODIUM BISULFATE AND 3 ml WHITE VINEGAR. HEAT IT GENTLY. SNIFF CAREFULLY. THE SOUR SMELL OF VINEGAR HAS TURNED INTO THE FRUITY SMELL OF ETHYL ACETATE ($CH_3COOC_2H_5$). IT IS A MUCH-USED SOLVENT.

CHLOROFORM FROM ETHANOL

MIX 5 ml ETHANOL WITH 5 ml SODIUM HYPOCHLORITE SOLUTION ("CLOROX"). HEAT MIXTURE GENTLY FOR A FEW MOMENTS WITHOUT BOILING. THEN SNIFF CAREFULLY. YOU GET THE PECULIAR SWEETISH ODOR OF CHLOROFORM. THE C_2H_5OH HAS BEEN TURNED INTO $CHCl_3$.

Carboxylic Acids

CAN YOU THINK of anything more refreshing than a glass of cold lemonade on a hot summer's day? Or anything better than cranberry sauce for adding a tangy taste to the Thanksgiving dinner?

The tartness of lemonade and cranberry sauce comes from organic acids.

These acids are found ready-made in nature in great numbers. Some of them occur as free acids (citric acid, tannic acid, malic acid), others as esters (products of acids and alcohols, such as fats and oils and the flavors of many fruits and the odors of many flowers). Still other of these organic acids are produced by the action of bacteria (acetic acid from wine or cider, lactic acid when milk turns sour, butyric acid in rancid butter).

Some organic acids can be extracted directly from the plant parts in which they are found. But to get them in pure and concentrated form it is usually necessary to turn them into sodium or calcium salts and then free the acids from the salts with a stronger acid. Many of the acids which were formerly obtained from plant parts can now be made artificially in the laboratory.

Organic acids have one thing in common. They all contain a combination of one carbon atom, one oxygen atom, and one hydroxyl group (OH). This COOH combination, called a carboxyl group (from a joining-up of the words *carb*on and hydr*oxyl*), has given the organic acids their scientific name, carboxylic acids. When these acids form salts it is the H in the carboxyl group that is replaced by a metal, as, for instance, when CH_3COOH (acetic acid) forms CH_3COONa (sodium acetate).

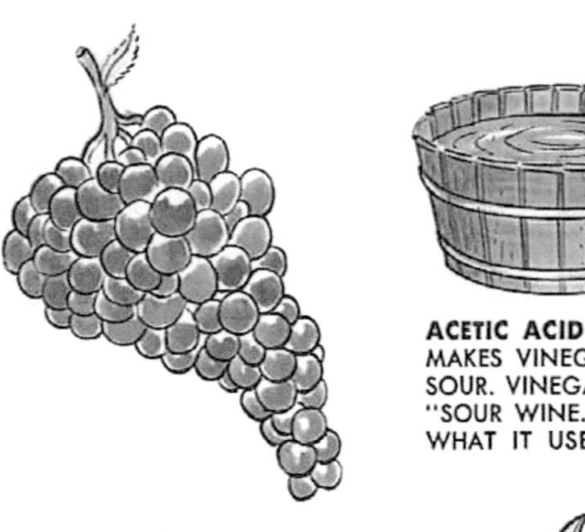

ACETIC ACID IS WHAT MAKES VINEGAR TASTE SOUR. VINEGAR MEANS "SOUR WINE." THAT IS WHAT IT USED TO BE.

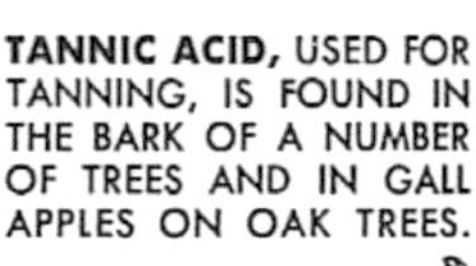

TANNIC ACID, USED FOR TANNING, IS FOUND IN THE BARK OF A NUMBER OF TREES AND IN GALL APPLES ON OAK TREES.

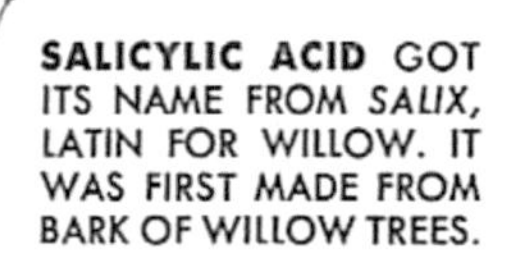

SALICYLIC ACID GOT ITS NAME FROM *SALIX,* LATIN FOR WILLOW. IT WAS FIRST MADE FROM BARK OF WILLOW TREES.

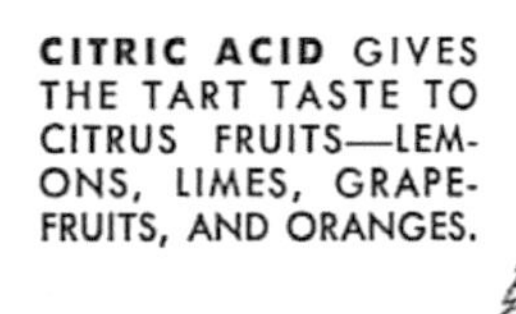

CITRIC ACID GIVES THE TART TASTE TO CITRUS FRUITS—LEMONS, LIMES, GRAPEFRUITS, AND ORANGES.

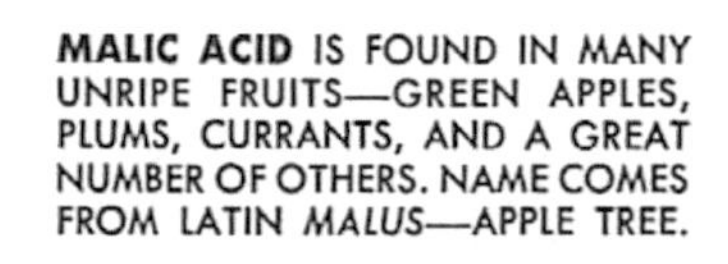

MALIC ACID IS FOUND IN MANY UNRIPE FRUITS—GREEN APPLES, PLUMS, CURRANTS, AND A GREAT NUMBER OF OTHERS. NAME COMES FROM LATIN *MALUS*—APPLE TREE.

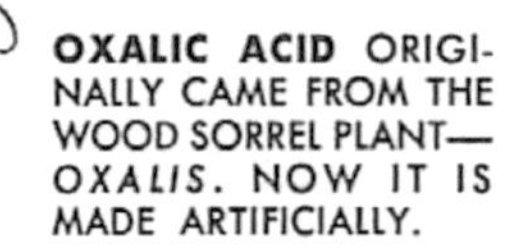

OXALIC ACID ORIGINALLY CAME FROM THE WOOD SORREL PLANT—*OXALIS.* NOW IT IS MADE ARTIFICIALLY.

FORMIC ACID IS THE HIGHLY IRRITATING ACID THAT ANTS *(FORMICA)* PUMP INTO YOU WHEN THEY BITE YOU.

ACETIC ACID

YOU HAVE ALREADY MADE SODIUM ACETATE (ON PAGE 83).

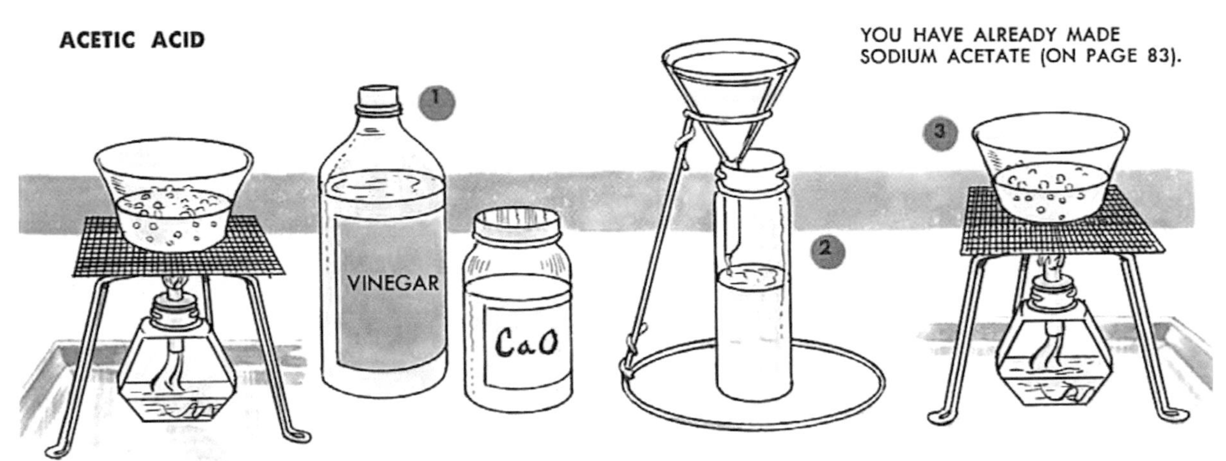

VINEGAR IS DILUTED ACETIC ACID. SEVERAL OF ITS SALTS—ACETATES—CAN BE MADE FROM VINEGAR. USE LIME FOR MAKING THE CALCIUM SALT—$(CH_3COO)_2Ca$.

1 WARM 50 ml WHITE VINEGAR IN A CUSTARD CUP. ADD CALCIUM OXIDE UNTIL NO MORE DISSOLVES.

2 FILTER SOLUTION TO REMOVE UNDISSOLVED CALCIUM OXIDE. FILTRATE CONTAINS CALCIUM ACETATE.

3 EVAPORATE SOLUTION UNTIL ALMOST DRY. DO NOT OVERHEAT—IF YOU DO, THE ACETATE BREAKS UP INTO CALCIUM CARBONATE AND ACETONE (CH_3COCH_3).

YOU CAN AGAIN DRIVE ACETIC ACID OUT OF ITS CALCIUM SALT.

MIX CALCIUM ACETATE WITH AN EQUAL AMOUNT OF SODIUM BISULFATE. PLACE IN DRY TEST TUBE. HEAT GENTLY. YOU GET SHARP ODOR OF ACETIC ACID. MOISTENED BLUE LITMUS PAPER AT MOUTH OF TUBE TURNS RED.

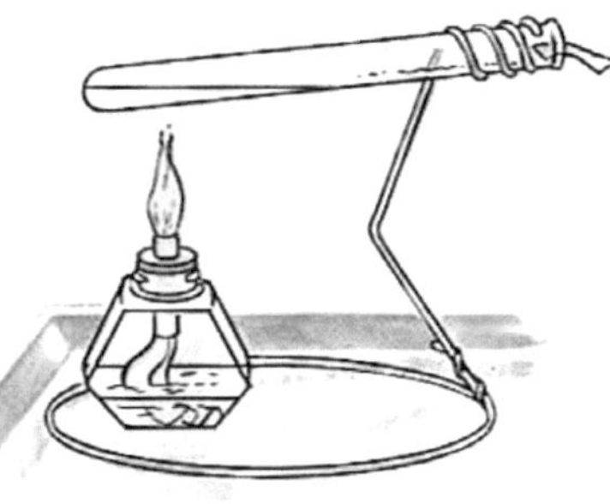

TANNIC ACID

TANNIC ACID IS FOUND IN TEA.

1 BOIL ¼ TEASPOON TEA IN 50 ml WATER. THEN LET IT STAND TO STEEP AND COOL. POUR OFF THE CLEAR LIQUID.

2 DISSOLVE A CRYSTAL OF IRON SULFATE IN 5 ml WATER AND ADD TO THE TEA. YOU WILL GET A BLACK PRECIPITATE OF IRON TANNATE.

SALICYLIC ACID

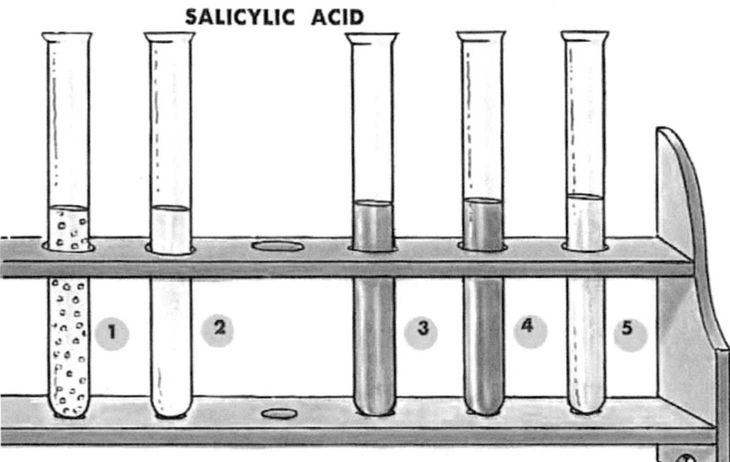

1 SHAKE UP 1 g SALICYLIC ACID WITH 10 ml WATER. IT DOES NOT GO INTO SOLUTION.

2 ADD 10 PER CENT NaOH SOLUTION BY THE DROP UNTIL ALL SALICYLIC ACID IS DISSOLVED. YOU NOW HAVE A SODIUM SALICYLATE SOLUTION.

3 WITH IRON SULFATE, SODIUM SALICYLATE GIVES RED-BROWN FERROUS SALICYLATE.

4 A FERRIC SALT GIVES WINE-RED FERRIC SALICYLATE.

5 COPPER SULFATE GIVES THE GREEN COPPER SALICYLATE.

PHENOL FROM SALICYLIC ACID

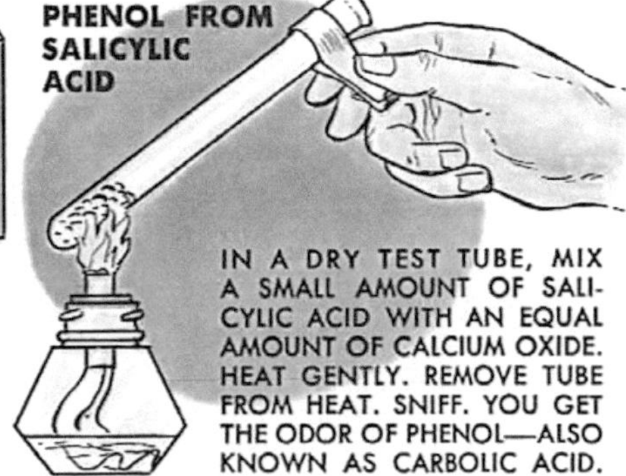

IN A DRY TEST TUBE, MIX A SMALL AMOUNT OF SALICYLIC ACID WITH AN EQUAL AMOUNT OF CALCIUM OXIDE. HEAT GENTLY. REMOVE TUBE FROM HEAT. SNIFF. YOU GET THE ODOR OF PHENOL—ALSO KNOWN AS CARBOLIC ACID.

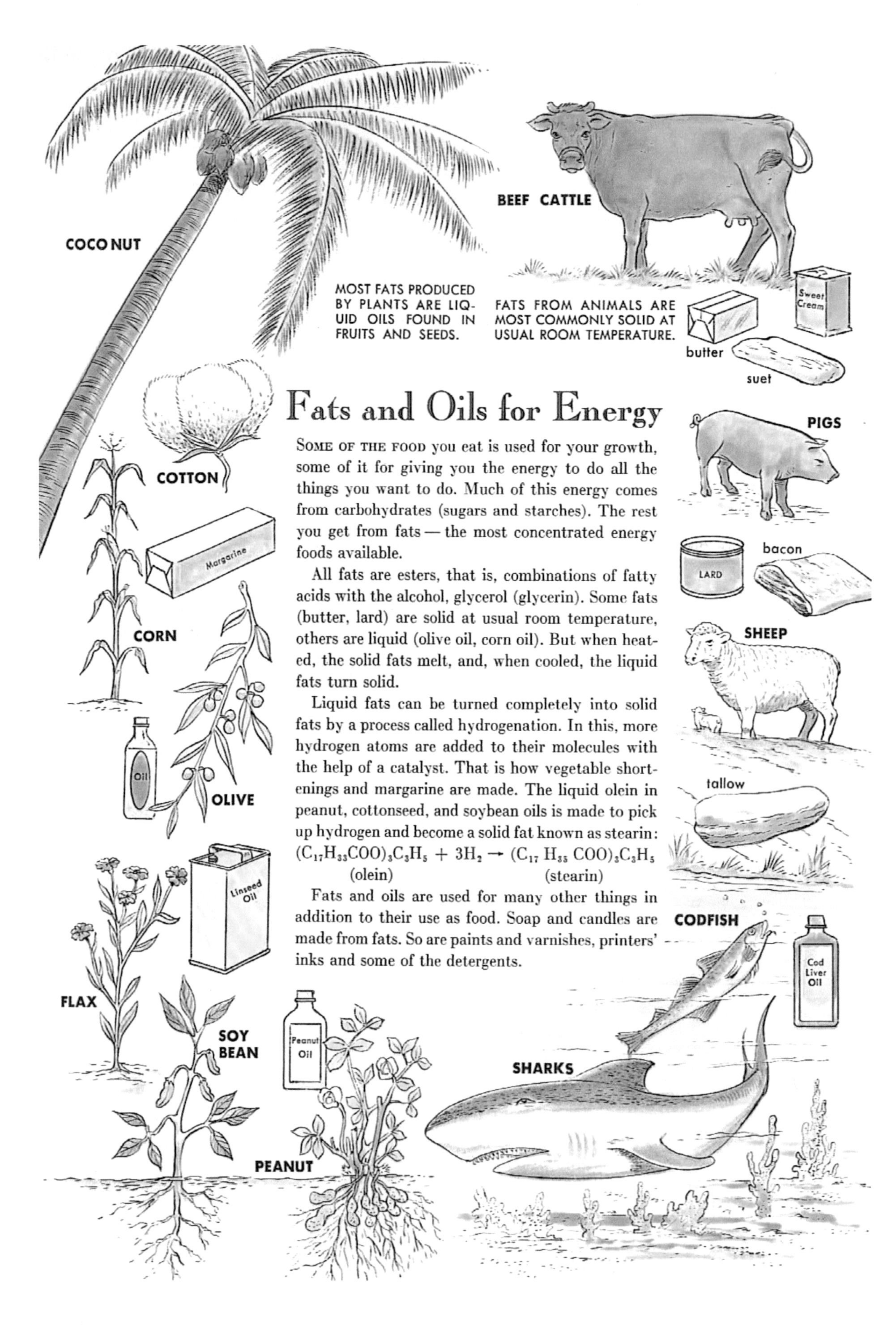

Fats and Oils for Energy

SOME OF THE FOOD you eat is used for your growth, some of it for giving you the energy to do all the things you want to do. Much of this energy comes from carbohydrates (sugars and starches). The rest you get from fats — the most concentrated energy foods available.

All fats are esters, that is, combinations of fatty acids with the alcohol, glycerol (glycerin). Some fats (butter, lard) are solid at usual room temperature, others are liquid (olive oil, corn oil). But when heated, the solid fats melt, and, when cooled, the liquid fats turn solid.

Liquid fats can be turned completely into solid fats by a process called hydrogenation. In this, more hydrogen atoms are added to their molecules with the help of a catalyst. That is how vegetable shortenings and margarine are made. The liquid olein in peanut, cottonseed, and soybean oils is made to pick up hydrogen and become a solid fat known as stearin:

$$\underset{\text{(olein)}}{(C_{17}H_{33}COO)_3C_3H_5} + 3H_2 \rightarrow \underset{\text{(stearin)}}{(C_{17}H_{35}COO)_3C_3H_5}$$

Fats and oils are used for many other things in addition to their use as food. Soap and candles are made from fats. So are paints and varnishes, printers' inks and some of the detergents.

EXTRACTING FAT

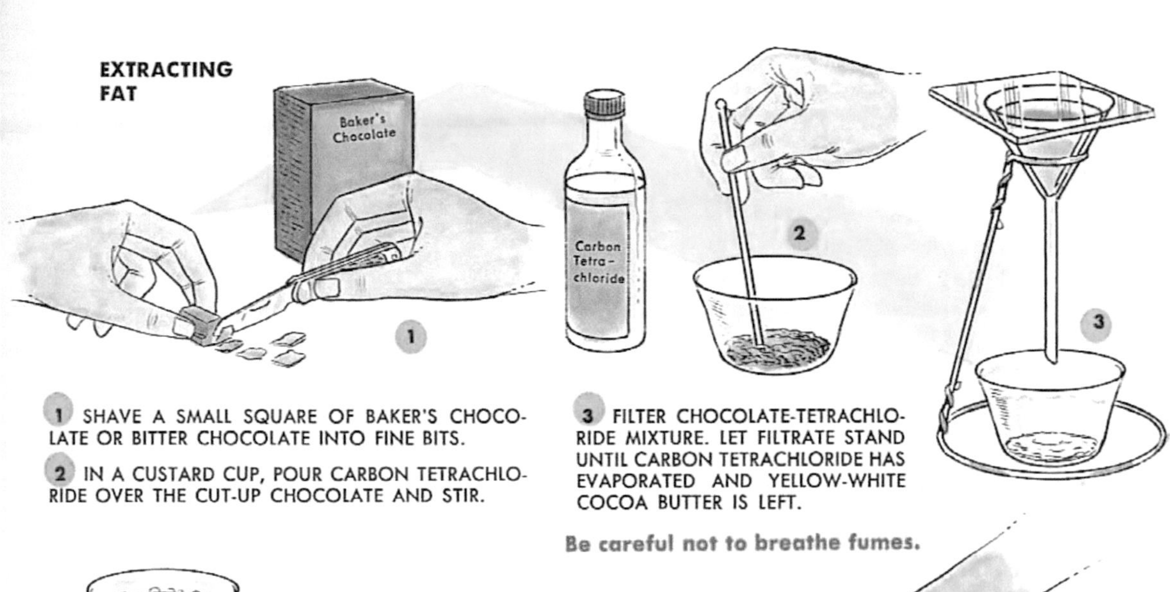

1 SHAVE A SMALL SQUARE OF BAKER'S CHOCOLATE OR BITTER CHOCOLATE INTO FINE BITS.

2 IN A CUSTARD CUP, POUR CARBON TETRACHLORIDE OVER THE CUT-UP CHOCOLATE AND STIR.

3 FILTER CHOCOLATE-TETRACHLORIDE MIXTURE. LET FILTRATE STAND UNTIL CARBON TETRACHLORIDE HAS EVAPORATED AND YELLOW-WHITE COCOA BUTTER IS LEFT.

Be careful not to breathe fumes.

RENDERING FAT

"RENDERING" IS THE MOST COMMON METHOD OF EXTRACTING FAT.

TEST FOR FAT

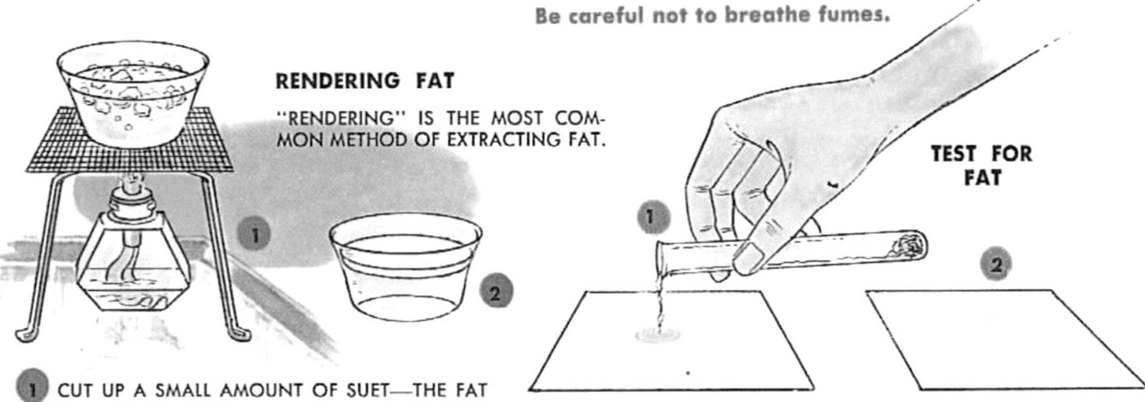

1 CUT UP A SMALL AMOUNT OF SUET—THE FAT FROM A PIECE OF BEEF. DROP IT INTO HOT WATER. BOIL WATER FOR TEN MINUTES OR MORE.

2 REMOVE THE RENDERED-OUT SUET. PLACE CUSTARD CUP IN REFRIGERATOR. AFTER COOLING YOU CAN LIFT OFF THE FAT AS A SOLID DISK.

1 CRUSH A COUPLE OF PEANUTS. DROP THEM IN A TEST TUBE. COVER THEM WITH CARBON TETRACHLORIDE AND LET STAND ABOUT 5 MINUTES. POUR A FEW DROPS ON A PIECE OF PAPER. LET CARBON TETRACHLORIDE EVAPORATE.

2 LOOK AT THE PAPER AGAINST THE LIGHT. THE ALMOST TRANSPARENT "GREASE SPOT" IS A TEST FOR FAT.

TEST FOR GLYCEROL (GLYCERIN)

IN A DRY TEST TUBE ADD 1/4 TEASPOON SODIUM BISULFATE TO 1 ml VEGETABLE OIL AND HEAT GENTLY. WAFT THE IRRITATING ODOR TOWARD YOU AND SNIFF CAUTIOUSLY. THE SMELL IS FROM ACROLEIN WHICH IS PRODUCED BY BREAKING DOWN THE GLYCERIN IN THE FAT.

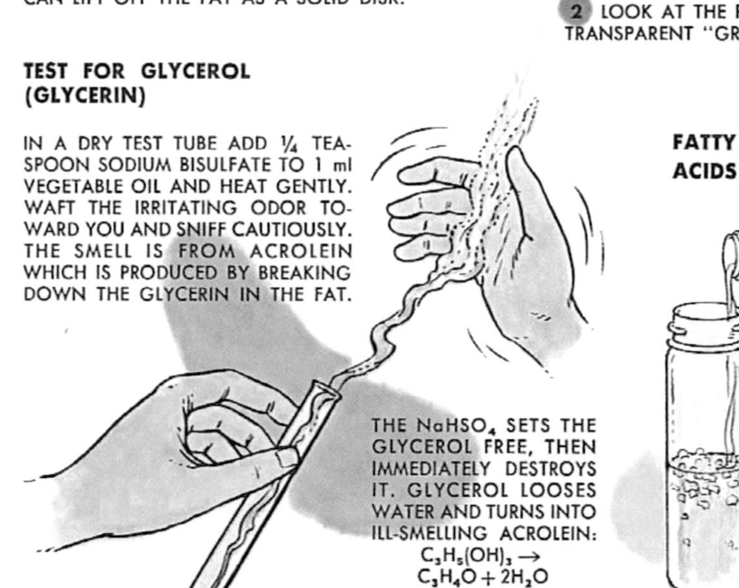

THE $NaHSO_4$ SETS THE GLYCEROL FREE, THEN IMMEDIATELY DESTROYS IT. GLYCEROL LOOSES WATER AND TURNS INTO ILL-SMELLING ACROLEIN:

$$C_3H_5(OH)_3 \rightarrow C_3H_4O + 2H_2O$$

FATTY ACIDS

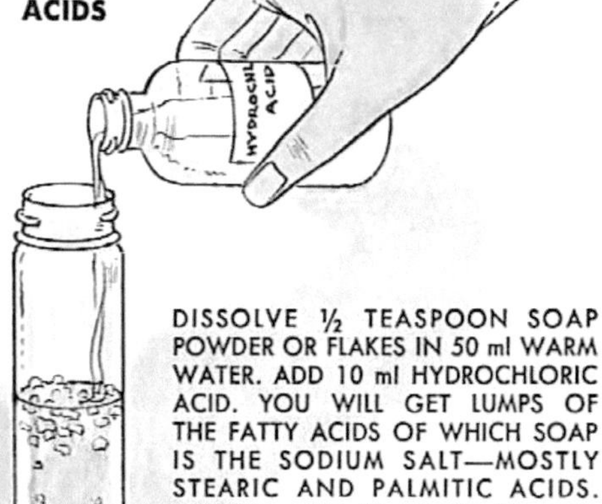

DISSOLVE 1/2 TEASPOON SOAP POWDER OR FLAKES IN 50 ml WARM WATER. ADD 10 ml HYDROCHLORIC ACID. YOU WILL GET LUMPS OF THE FATTY ACIDS OF WHICH SOAP IS THE SODIUM SALT—MOSTLY STEARIC AND PALMITIC ACIDS. STEARIC ACID IS ADDED TO PARAFFIN IN THE MAKING OF CANDLES.

IN THE OLD-FASHIONED SOAP KETTLE, ONLY A FEW GALLONS OF SOAP COULD BE MADE AT ONE TIME.

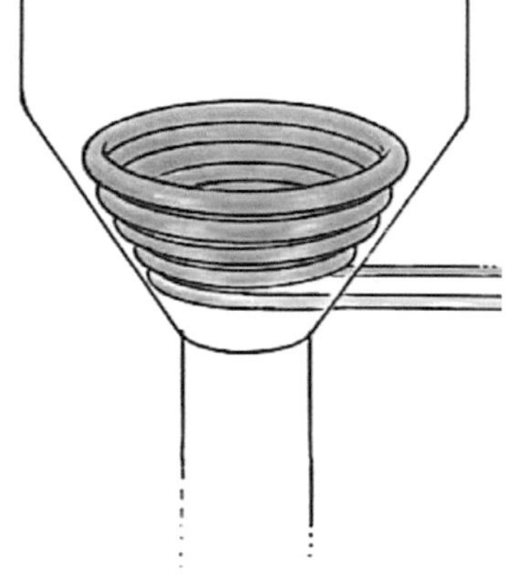

IN MODERN SOAP PANS, SEVERAL STORIES HIGH, UP TO 100 TONS OF FAT CAN BE TURNED INTO SOAP.

Soap and Soap Making

WHENEVER YOUR HANDS get dirty, it is an easy matter to get them clean. All you need is water and plenty of $CH_3CH_2CH_2CH_2CH_2CH_2CH_2CH_2CH_2CH_2CH_2CH_2CH_2CH_2CH_2CH_2CH_2COONa$ — $C_{17}H_{35}COONa$ for short, the sodium salt of stearic acid, a substance more generally known as soap.

Soap has been used for cleaning for thousands of years. No one knows who invented it — but the method for making it was passed down from father to son, from mother to daughter. The early soap makers first had to burn wood to get potash (K_2CO_3—see page 59) or dried seaweed to get soda ash (Na_2CO_3). This was treated with lime to make potassium or sodium hydroxide (KOH or $NaOH$ — see page 45), and this, in turn, was boiled with fat to make soap. Very much the same method is used today — except that the boiling is done in tremendous soap pans under steam pressure.

STRONG SOAP BUBBLES RESULT WHEN YOU ADD GLYCERIN TO THE SOAP SOLUTION. HERE IS A RECIPE: 5 g SOAP, 100 ml WATER, AND 10 ml GLYCERIN.

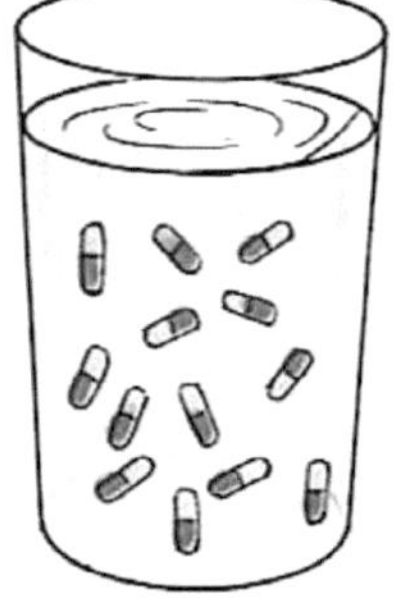

THIS IS THE WAY SCIENTISTS BELIEVE THAT SOAP ACTS: ONE END OF THE SOAP MOLECULE IS SOLUBLE IN WATER, THE OTHER END IN OIL. WHEN OIL IS SHAKEN UP IN SOAPY WATER, THE OIL DROPS ARE SURROUNDED BY THE SOAP MOLECULES DIPPING THE OIL-DISSOLVING ENDS INTO THE OIL. THE WATER-SOLUBLE ENDS HOLD THE OIL DROPLETS SUSPENDED.

MAKING SOAP

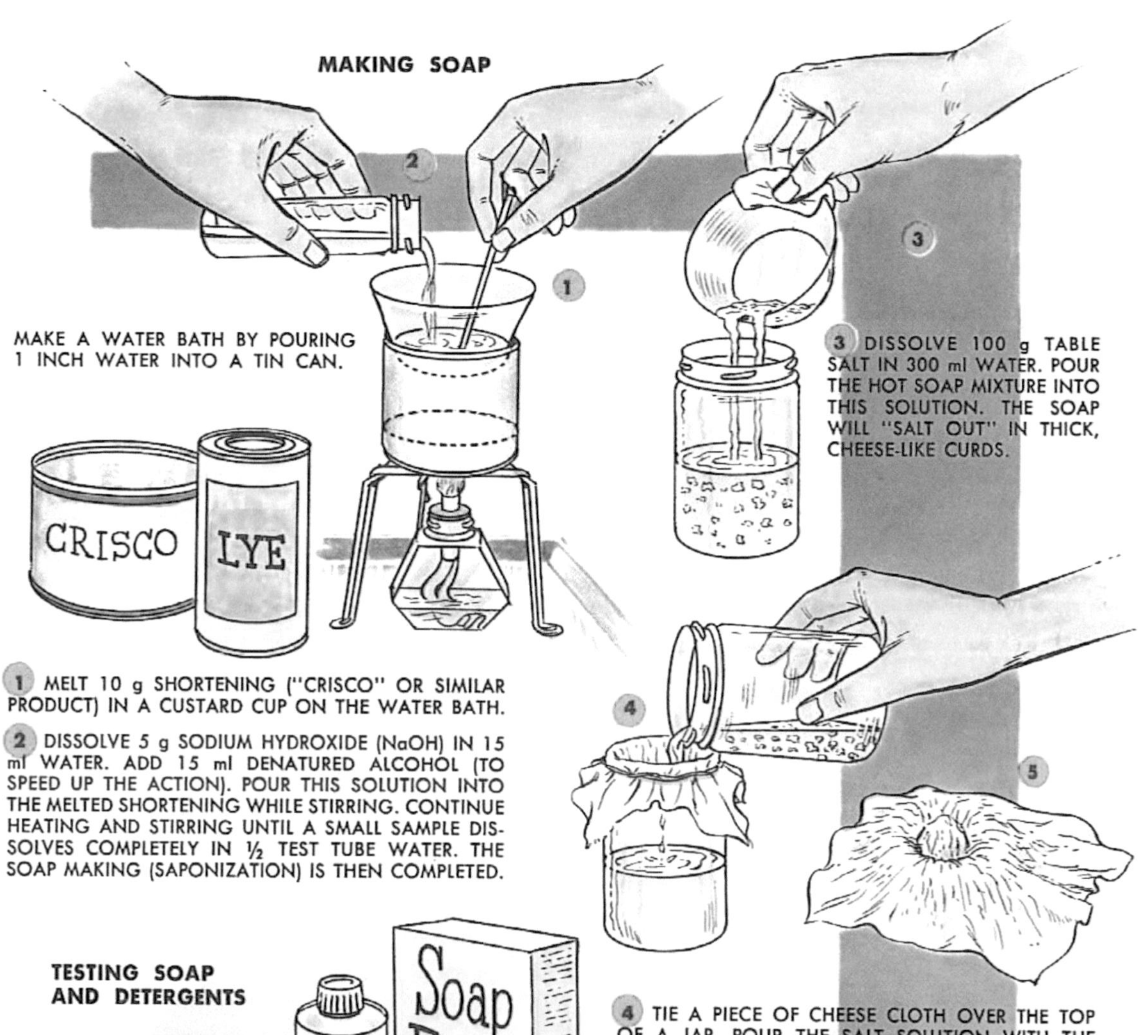

MAKE A WATER BATH BY POURING 1 INCH WATER INTO A TIN CAN.

1 MELT 10 g SHORTENING ("CRISCO" OR SIMILAR PRODUCT) IN A CUSTARD CUP ON THE WATER BATH.

2 DISSOLVE 5 g SODIUM HYDROXIDE ($NaOH$) IN 15 ml WATER. ADD 15 ml DENATURED ALCOHOL (TO SPEED UP THE ACTION). POUR THIS SOLUTION INTO THE MELTED SHORTENING WHILE STIRRING. CONTINUE HEATING AND STIRRING UNTIL A SMALL SAMPLE DISSOLVES COMPLETELY IN ½ TEST TUBE WATER. THE SOAP MAKING (SAPONIZATION) IS THEN COMPLETED.

3 DISSOLVE 100 g TABLE SALT IN 300 ml WATER. POUR THE HOT SOAP MIXTURE INTO THIS SOLUTION. THE SOAP WILL "SALT OUT" IN THICK, CHEESE-LIKE CURDS.

4 TIE A PIECE OF CHEESE CLOTH OVER THE TOP OF A JAR. POUR THE SALT SOLUTION WITH THE SOAP CURDS INTO CHEESE CLOTH AND LET SALT SOLUTION DRAIN OFF. WASH THE SOAP BY POURING TWO TEST TUBES OF ICE-COLD WATER THROUGH IT TO REMOVE MOST OF THE SALT THAT'S STILL ON IT.

5 FINALLY, SQUEEZE OUT THE WATER AND SPREAD OUT THE CHEESE CLOTH TO LET THE SOAP DRY.

TESTING SOAP AND DETERGENTS

DISSOLVE 1 g OF YOUR HOME-MADE SOAP IN 50 ml LUKEWARM WATER. ALSO MAKE SOLUTIONS IN 50 ml WATER OF 1 g TOILET SOAP, 1 g SOAP FLAKES, 1 g SOAP POWDER, 1 g POWDERED DETERGENT, AND 1 ml LIQUID DETERGENT.

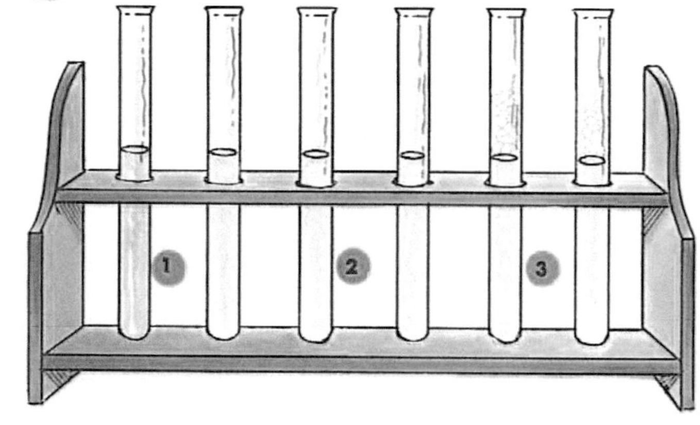

1 POUR 10 ml OF THE SOAP AND DETERGENT SOLUTIONS INTO SEPARATE TEST TUBES. TEST EACH SOLUTION FOR ACID AND BASE WITH LITMUS PAPER AND PHENOLPHTHALEIN.

2 SHAKE 5 DROPS OF OIL INTO EACH SOLUTION. NOTE THE DIFFERENCE IN THE WAY THE SOLUTIONS MAKE EMULSION WITH OIL.

3 AGAIN, POUR 10 ml OF EACH SOLUTION INTO SEPARATE TEST TUBES. ADD 5 ml LIMEWATER TO EACH. SHAKE AND NOTICE THE DIFFERENCE IN THE AMOUNT OF FOAM MADE BY EACH SOLUTION IN THIS "HARD" WATER.

Proteins—the Body-Building Foods

At almost every meal, we look forward especially to the proteins: ham and eggs for breakfast, hamburgers or frankfurters for lunch, steak or chicken for dinner. We drink milk mostly for the sake of its proteins. Even many of our desserts are protein products — from ice cream to Jell-O.

While most other foodstuffs, such as carbohydrates and fats, consist of carbon, hydrogen, and oxygen, the proteins also contain nitrogen and, for the most part, sulfur. Their molecules are "giants" compared with the molecules of other chemical compounds. One of them, albumin in egg, has this estimated formula: $C_{696}H_{1125}O_{200}N_{190}S_{18}$.

Not all proteins are edible. You would hardly think of eating hair and nails, furs and feathers — yet these are all proteins. (CONTINUED ON PAGE 99)

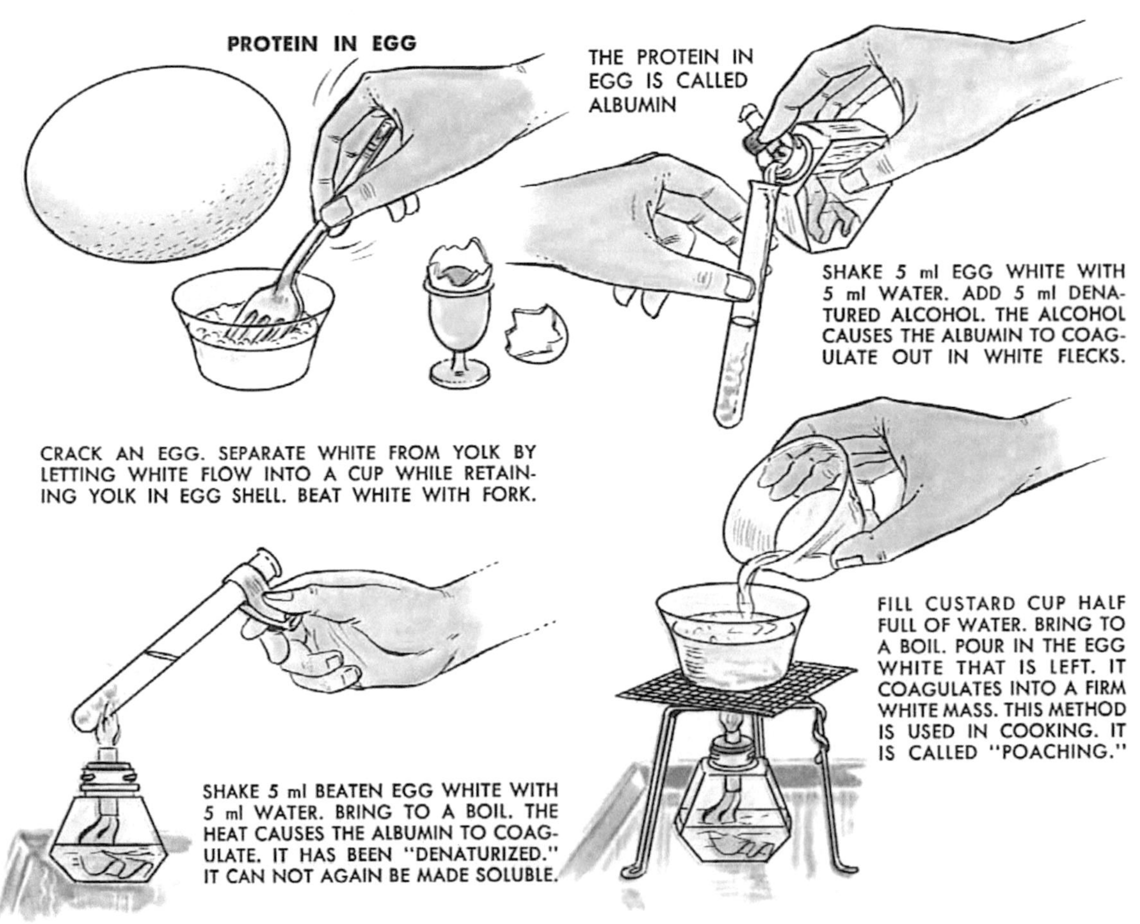

THE ITEMS ON THE TOP OF THESE TWO PAGES ALL CONTAIN PROTEINS.

WHAT DOES ALBUMIN CONSIST OF?

1 DROP A SMALL PIECE OF COAGULATED EGG WHITE INTO A TEST TUBE. COVER IT WITH 5 ml 10% NaOH SOLUTION. HEAT. WHITE GOES IN SOLUTION.

2 POUR A FEW DROPS OF THE EGG WHITE SOLUTION ONTO A BRIGHT SILVER COIN. IN A FEW MINUTES SILVER COIN TURNS BROWNISH-BLACK FROM SILVER SULFIDE, PROVING THAT ALBUMIN CONTAINS SULFUR.

1 PLACE A SMALL PIECE OF COAGULATED EGG WHITE ON A PIECE OF TIN. HEAT. VAPORS SMELL OF AMMONIA AND TURN WETTED RED LITMUS PAPER BLUE. AMMONIA IS NH_3. ALBUMIN MUST CONTAIN N AND H.

ALBUMIN IS FOUND IN EGGS, BLOOD, MILK, AND GRAIN.

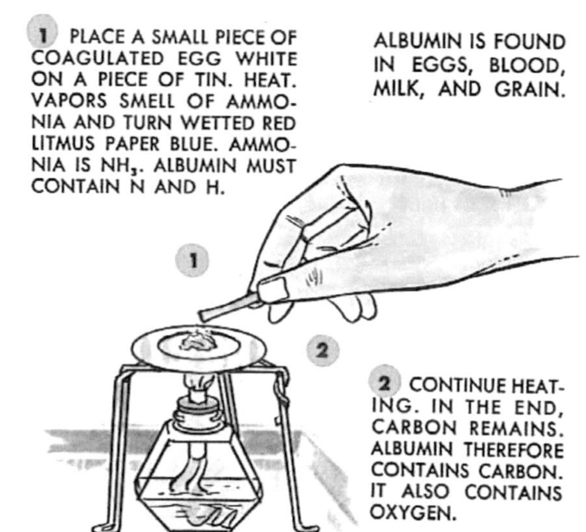

2 CONTINUE HEATING. IN THE END, CARBON REMAINS. ALBUMIN THEREFORE CONTAINS CARBON. IT ALSO CONTAINS OXYGEN.

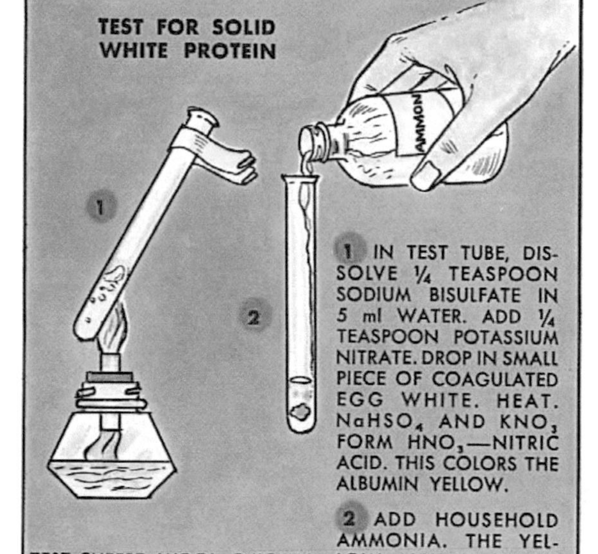

TEST FOR SOLID WHITE PROTEIN

1 IN TEST TUBE, DISSOLVE ¼ TEASPOON SODIUM BISULFATE IN 5 ml WATER. ADD ¼ TEASPOON POTASSIUM NITRATE. DROP IN SMALL PIECE OF COAGULATED EGG WHITE. HEAT. $NaHSO_4$ AND KNO_3 FORM HNO_3—NITRIC ACID. THIS COLORS THE ALBUMIN YELLOW.

2 ADD HOUSEHOLD AMMONIA. THE YELLOW ALBUMIN TURNS BRIGHT ORANGE.

TEST CHEESE, WOOL, CHICKEN, LIMA BEANS THE SAME WAY.

WHAT DOES EGG YOLK CONTAIN?

Be careful not to breathe fumes.

1 SHAKE 5 ml OF THE YOLK WITH 5 ml CARBON TETRACHLORIDE TO FIND OUT IF IT CONTAINS FAT.

2 POUR A LITTLE OUT ON PAPER. LET CARBON TETRACHLORIDE EVAPORATE. GREASE SPOT REMAINS.

3 HEAT THE MIXTURE. YOU GET A WHITE COAGULATION. YOLK AND WHITE BOTH CONTAIN ALBUMIN.

PROTEIN IN MILK

MILK IS AN IMPORTANT SOURCE OF PROTEIN. THE PROTEIN IN MILK IS CALLED CASEIN. CHEESE IS SPECIALLY TREATED CASEIN.

1 POUR ½ CUP SKIM MILK (OR MIXTURE OF 8 TEASPOONS SKIM MILK POWDER AND ½ CUP WATER) INTO A CUSTARD CUP. HEAT GENTLY UNTIL IT FEELS JUST SLIGHTLY WARM WHEN YOU TEST IT WITH A FINGER.

2 ADD ONE TEST TUBE FULL OF WHITE VINEGAR TO THE WARM SKIM MILK. THE CASEIN SEPARATES IN HEAVY, WHITE CURDS.

3 TIE A PIECE OF CHEESE CLOTH OVER A JAR. POUR THE CURDLED MILK INTO THE CHEESE CLOTH. LET LIQUID (WHEY MIXED WITH VINEGAR) RUN OUT. KEEP THE LIQUID.

4 FOLD CHEESE CLOTH UP AROUND THE CASEIN. DIP THE BAG IN WATER AND SQUEEZE SEVERAL TIMES TO WASH OUT WHEY AND VINEGAR.

5 SQUEEZE THE CASEIN ALMOST DRY. SPREAD OUT THE CHEESE CLOTH TO LET THE CASEIN DRY.

WHAT ELSE IS IN MILK?

1 POUR THE VINEGAR-MIXED WHEY INTO A CUSTARD CUP AND BRING IT TO A BOIL. YOU WILL SEE TINY WHITE FLECKS. THESE ARE ALBUMIN COAGULATED OUT BY THE HEAT.

2 FILTER THE WHEY. TEST THE FILTRATE WITH FEHLING SOLUTION (SEE PAGE 85). MILK SUGAR GIVES RED Cu_2O PRECIPITATE.

MAKING CASEIN GLUE

SOFTEN 4 g CASEIN WITH 4 ml WATER. SHAKE UP 1 g CALCIUM OXIDE IN 4 ml WATER. POUR THE CALCIUM OXIDE MIXTURE INTO THE CASEIN WHILE STIRRING. THE RESULTING SMOOTH PASTE IS AN EXCELLENT GLUE FOR PAPER AND FOR WOOD.

GELATIN IS A PROTEIN

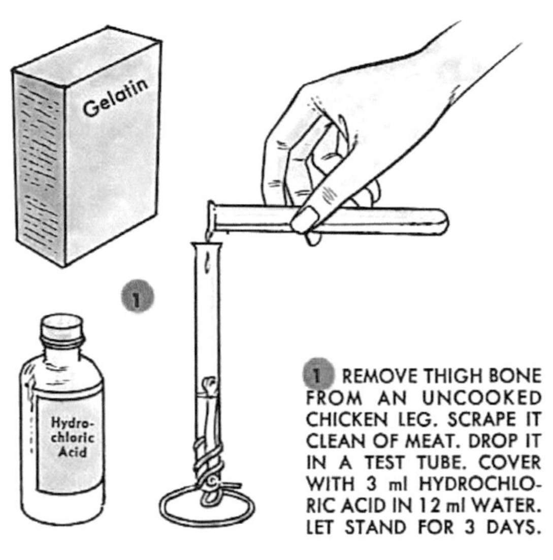

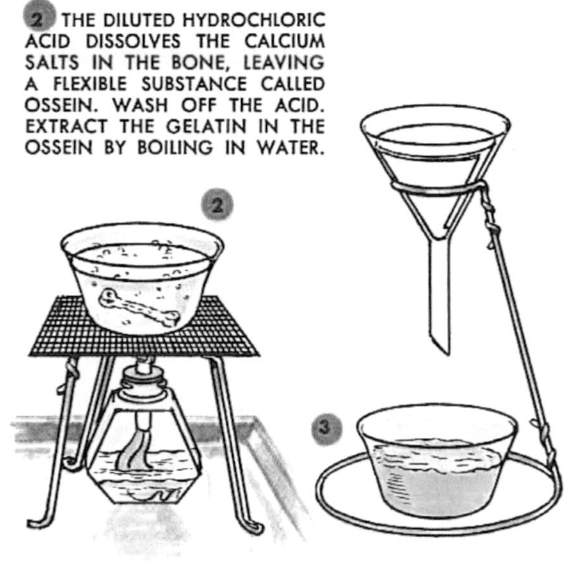

Proteins—Continued

You are certain to be familiar with three common, pure proteins: albumin in eggs, casein in milk, and gelatin.

ALBUMIN — Egg white contains around 13 per cent albumin — from Latin *albus*, white.

When you shake up egg white with water, you get what looks like an almost clear solution. But this is not a "true" solution such as you get when you dissolve salt or sugar — it is another kind of "solution" called a "colloidal dispersion." For more about colloidal dispersions, see pages 100-101.

As long as egg white is kept cool, it stays transparent and almost liquid. But what happens when you heat it? You know from frying or boiling an egg: It hardens — coagulates — into a solid white mass which you can not again "dissolve" in water. The chemist's term for this change is "denaturation" — the egg white has changed its nature.

CASEIN — Casein is another protein that goes into your diet. Some of the casein you drink (milk), some of it you eat (ice cream and cheese).

In cheese making, the casein is separated from the liquid part of the milk — the whey. It is then pressed and stored until ripe. The flavors of cheeses are caused mostly by esters created during the ripening.

GELATIN — Gelatin is a protein made from animal skins and bones, horns and hooves.

Gelatin behaves in a peculiar manner with water. In cold water it merely swells, but in hot water it "dissolves" readily, forming a colloidal dispersion. As long as you keep this dispersion warm, it remains in a liquid form that is called a "sol." When cooled, it turns into a jelly-like form called a "gel."

TEST FOR LIQUID PROTEINS

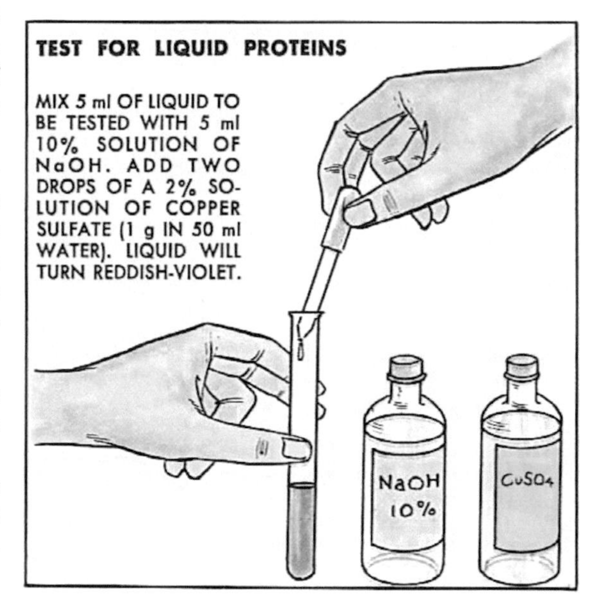

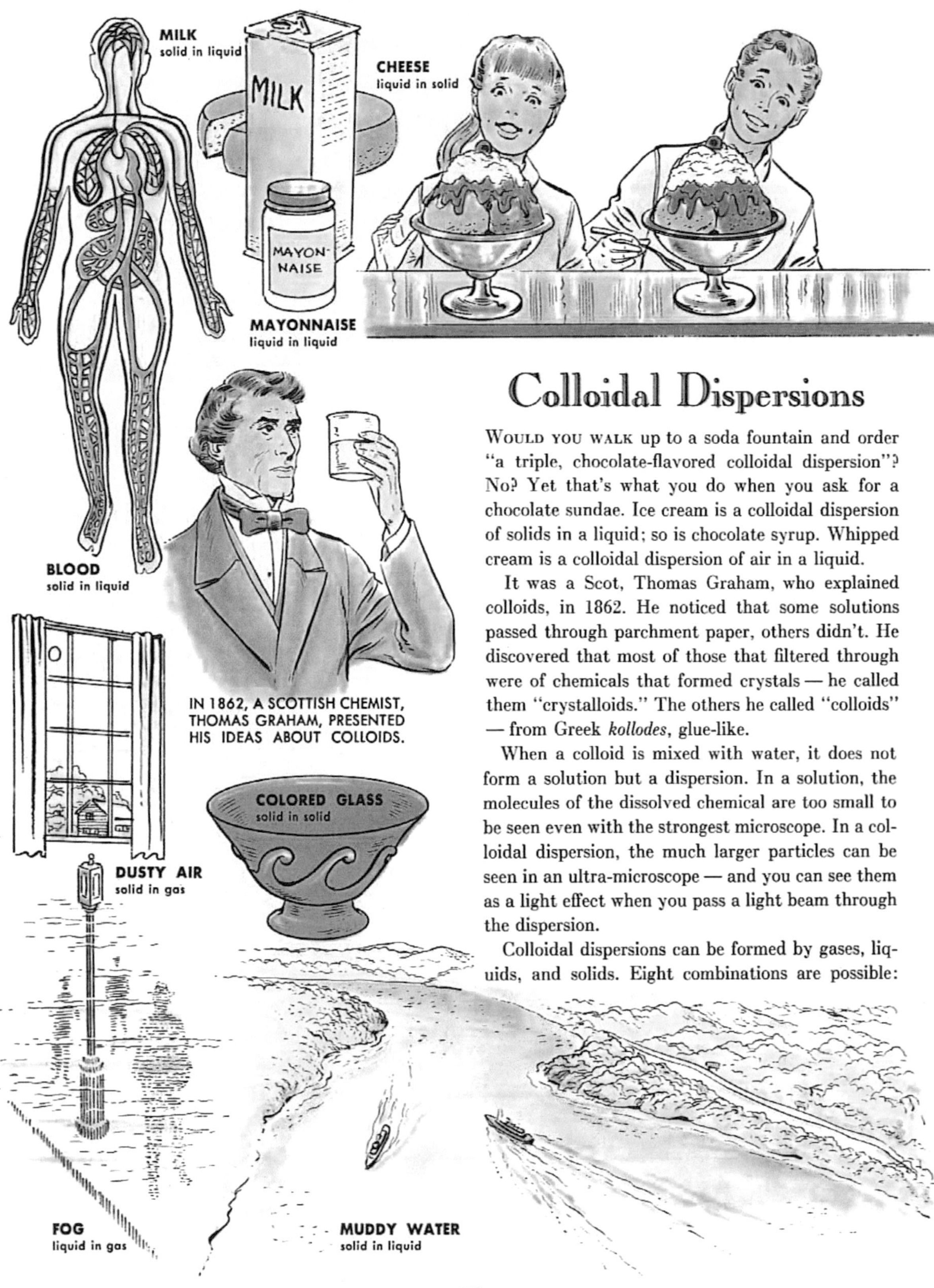

IN 1862, A SCOTTISH CHEMIST, THOMAS GRAHAM, PRESENTED HIS IDEAS ABOUT COLLOIDS.

Colloidal Dispersions

WOULD YOU WALK up to a soda fountain and order "a triple, chocolate-flavored colloidal dispersion"? No? Yet that's what you do when you ask for a chocolate sundae. Ice cream is a colloidal dispersion of solids in a liquid; so is chocolate syrup. Whipped cream is a colloidal dispersion of air in a liquid.

It was a Scot, Thomas Graham, who explained colloids, in 1862. He noticed that some solutions passed through parchment paper, others didn't. He discovered that most of those that filtered through were of chemicals that formed crystals — he called them "crystalloids." The others he called "colloids" — from Greek *kollodes*, glue-like.

When a colloid is mixed with water, it does not form a solution but a dispersion. In a solution, the molecules of the dissolved chemical are too small to be seen even with the strongest microscope. In a colloidal dispersion, the much larger particles can be seen in an ultra-microscope — and you can see them as a light effect when you pass a light beam through the dispersion.

Colloidal dispersions can be formed by gases, liquids, and solids. Eight combinations are possible:

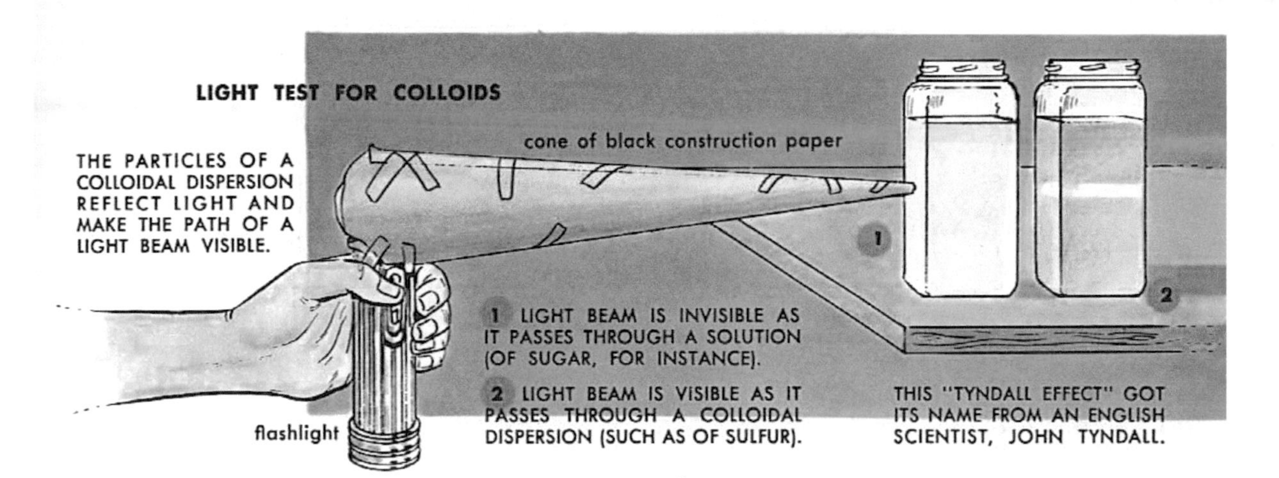

gases in liquids and in solids; liquids in gases, in other liquids, and in solids; solids in gases, in liquids, and in other solids. The illustrations show some of these possibilities — you can think of many others.

The colloidal state is important to life. It is the way in which we get most of our food, the way we digest it, and the way the blood carries nourishment throughout our bodies.

IN **PEPTIZATION,** LARGE PARTICLES ARE BROKEN DOWN INTO SMALLER PARTICLES OF COLLOIDAL SIZE.

1 SHAKE UP 1 g STARCH WITH 100 ml COLD WATER. IF LEFT UNDISTURBED, STARCH QUICKLY SETTLES TO BOTTOM.

2 POUR THE MIXTURE OF STARCH AND WATER INTO A CUSTARD CUP. BRING TO A BOIL, THEN COOL. STARCH HAS NOW FORMED A COLLOIDAL DISPERSION.

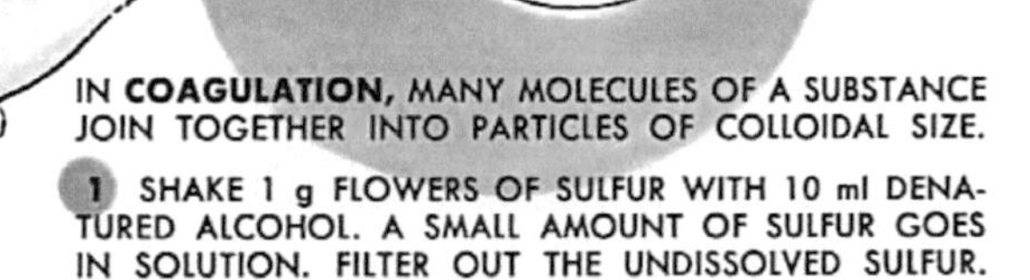

IN **COAGULATION,** MANY MOLECULES OF A SUBSTANCE JOIN TOGETHER INTO PARTICLES OF COLLOIDAL SIZE.

1 SHAKE 1 g FLOWERS OF SULFUR WITH 10 ml DENATURED ALCOHOL. A SMALL AMOUNT OF SULFUR GOES IN SOLUTION. FILTER OUT THE UNDISSOLVED SULFUR.

2 POUR THE ALCOHOLIC SOLUTION OF SULFUR INTO A LARGE AMOUNT OF WATER. YOU WILL SEE A WHITE CLOUD OF FINELY DISPERSED COLLOIDAL SULFUR.

IN **EMULSIFICATION,** ONE LIQUID IS DISPERSED IN ANOTHER. EMULSIONS CAN BE TEMPORARY OR PERMANENT.

1 SHAKE 5 ml KEROSENE AND 5 ml WATER TOGETHER IN A TEST TUBE. LET STAND FOR A SHORT TIME. LIQUIDS SEPARATE. THE EMULSION WAS TEMPORARY.

2 SHAKE 5 ml KEROSENE WITH SOLUTION OF ½ g SOAP IN 5 ml WARM WATER. THEN LET STAND. LIQUIDS DO NOT SEPARATE. THIS IS A PERMANENT EMULSION.

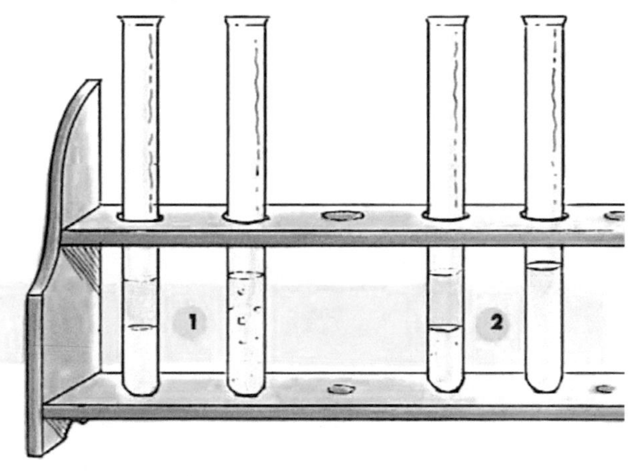

Natural and Artificial Fibers

IT WOULD BE TOUGH to get along without fibers in the modern world. Fibers are spun into thread, and the thread is made into cloth for clothing and bedsheets, curtains and towels, and many other things around the house. Fibers also go into such articles as string and rope, rugs and auto tires. Some of these fibers come from the plant and animal worlds, others are manufactured synthetically with coal or petroleum for their starting point.

Fibers belong in different groups of chemical compounds. Animal fibers are proteins; vegetable fibers are cellulose. Artificial fibers such as nylon, Orlon and Dacron are very complex chemical compounds and have enormously long molecules.

BURNING TEST FOR FIBERS

CUT HALF-INCH STRIPS OF DIFFERENT FABRICS. IGNITE EACH STRIP IN TURN. NOTICE HOW FABRIC BURNS, THE SMELL, AND ASH LEFT BEHIND.

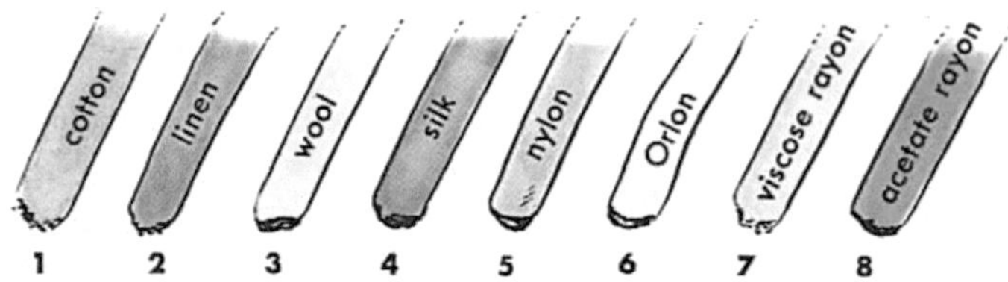

KIND	FLAME	SMELL	ASH
1 **COTTON**	Rapid, yellow flame	Like burning paper	Small, fine, gray
2 **LINEN**	Fairly fast, yellow flame	Like cotton	Like cotton
3 **WOOL**	Slow, sizzling flame	Like burning hair	Hollow, black bead, easy to crush
4 **SILK**	Small, slow flame	Like wool	Shiny, round bead, easy to crush
5 **NYLON**	Melts; no flame	Like celery	Melts to black bead, hard to crush
6 **ORLON**	Melts and burns	Like broiled fish	Black bead, hard to crush
7 **VISCOSE RAYON**	Rapid, yellow flame	Like cotton	Like cotton
8 **CELLULOSE ACETATE**	Rapid flame with small sparks; melts	Like vinegar	Black bead, hard to crush

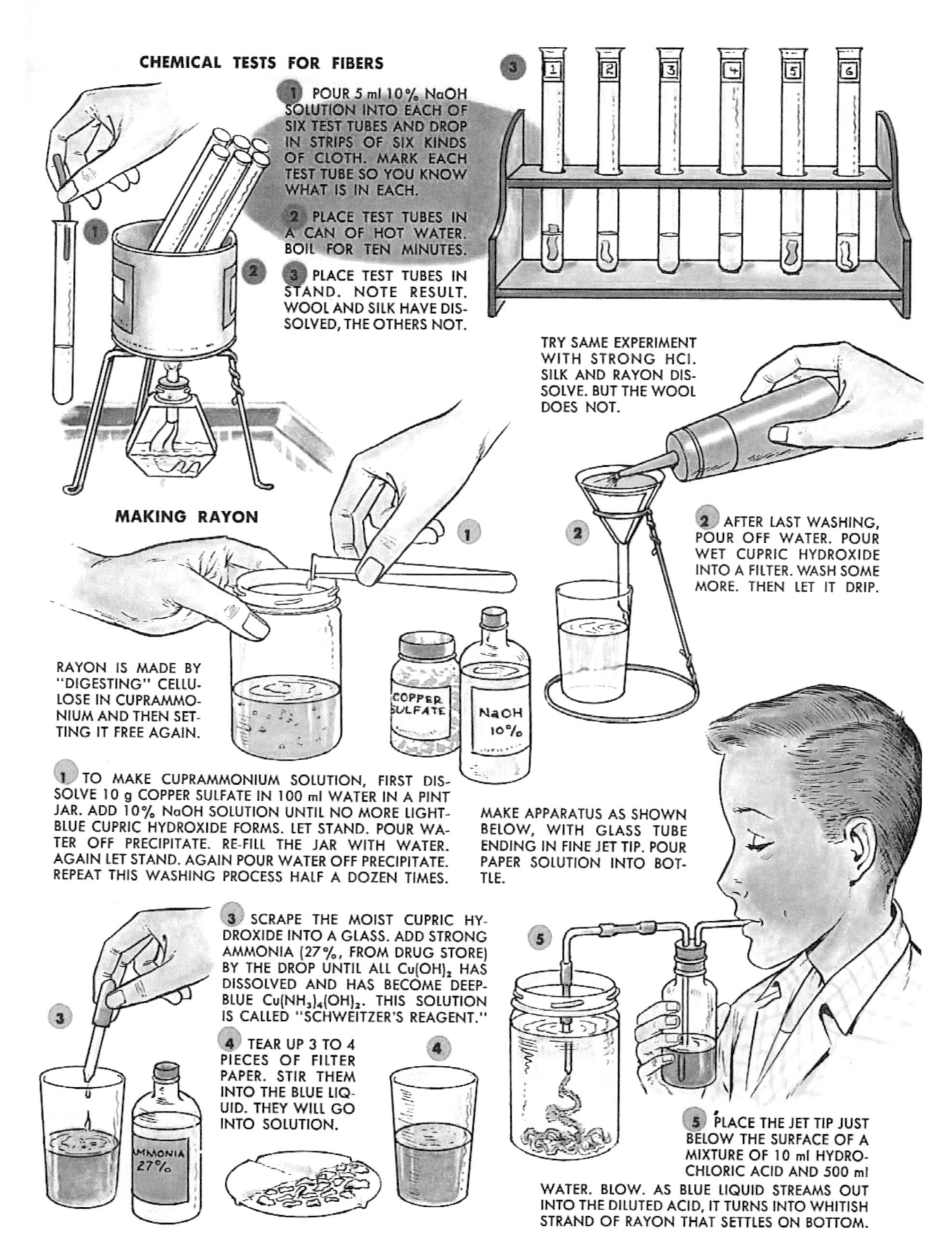

CHEMICAL TESTS FOR FIBERS

1 POUR 5 ml 10% NaOH SOLUTION INTO EACH OF SIX TEST TUBES AND DROP IN STRIPS OF SIX KINDS OF CLOTH. MARK EACH TEST TUBE SO YOU KNOW WHAT IS IN EACH.

2 PLACE TEST TUBES IN A CAN OF HOT WATER. BOIL FOR TEN MINUTES.

3 PLACE TEST TUBES IN STAND. NOTE RESULT. WOOL AND SILK HAVE DISSOLVED, THE OTHERS NOT.

TRY SAME EXPERIMENT WITH STRONG HCl. SILK AND RAYON DISSOLVE. BUT THE WOOL DOES NOT.

MAKING RAYON

RAYON IS MADE BY "DIGESTING" CELLULOSE IN CUPRAMMONIUM AND THEN SETTING IT FREE AGAIN.

1 TO MAKE CUPRAMMONIUM SOLUTION, FIRST DISSOLVE 10 g COPPER SULFATE IN 100 ml WATER IN A PINT JAR. ADD 10% NaOH SOLUTION UNTIL NO MORE LIGHT-BLUE CUPRIC HYDROXIDE FORMS. LET STAND. POUR WATER OFF PRECIPITATE. RE-FILL THE JAR WITH WATER. AGAIN LET STAND. AGAIN POUR WATER OFF PRECIPITATE. REPEAT THIS WASHING PROCESS HALF A DOZEN TIMES.

2 AFTER LAST WASHING, POUR OFF WATER. POUR WET CUPRIC HYDROXIDE INTO A FILTER. WASH SOME MORE. THEN LET IT DRIP.

3 SCRAPE THE MOIST CUPRIC HYDROXIDE INTO A GLASS. ADD STRONG AMMONIA (27%, FROM DRUG STORE) BY THE DROP UNTIL ALL $Cu(OH)_2$ HAS DISSOLVED AND HAS BECOME DEEP-BLUE $Cu(NH_3)_4(OH)_2$. THIS SOLUTION IS CALLED "SCHWEITZER'S REAGENT."

4 TEAR UP 3 TO 4 PIECES OF FILTER PAPER. STIR THEM INTO THE BLUE LIQUID. THEY WILL GO INTO SOLUTION.

MAKE APPARATUS AS SHOWN BELOW, WITH GLASS TUBE ENDING IN FINE JET TIP. POUR PAPER SOLUTION INTO BOTTLE.

5 PLACE THE JET TIP JUST BELOW THE SURFACE OF A MIXTURE OF 10 ml HYDROCHLORIC ACID AND 500 ml WATER. BLOW. AS BLUE LIQUID STREAMS OUT INTO THE DILUTED ACID, IT TURNS INTO WHITISH STRAND OF RAYON THAT SETTLES ON BOTTOM.

LEO H. BAEKELAND WITH THE AUTOCLAVE IN WHICH HE MADE BAKELITE IN HIS YONKERS, N. Y., LABORATORY.

Plastics—a Modern Giant

About fifty years ago, Dr. Leo H. Baekeland, a Belgian-born American chemist, mixed phenol and formaldehyde together during an experiment. Other chemists had done this before Baekeland and had wondered how to get the messy goo that resulted out of their test tubes. But Baekeland had another approach. He asked himself, "What is it good for?" He decided to find out. The result was Bakelite — the first successful, modern plastic.

During the year 1910, Baekeland produced less than 25 barrels of his "phenolic" plastic in a barn in Yonkers, N. Y. Nowadays, fifty years later, close to 500 million pounds are produced yearly. During those same fifty years, more than a dozen other types of plastics were invented.

Today, plastics seem to be everywhere. You find them in your home in flooring and wall coverings, in table tops and chair upholstery, in TV cabinets and telephones, in toys and games, in rigid containers and in squeeze bottles. Much of your food comes to you protected by some kind of plastic. They are used in planes and trains and cars. A plastic puts the "safety" into safety glass. Other plastics are used for long-wearing engine parts and for electrical insulation.

Plastics are made from a few simple raw materials — some just from water, air, and coal, others with the help of petroleum or natural gas, limestone and salt. The plastics chemist breaks down the comparatively simple molecules of these materials, then builds them up anew into very complex molecules.

Plastics may be divided into two main groups according to their special properties. One group consists of the thermosetting plastics. These can be molded by heat and pressure, but can not be remelted and remolded. They are along the lines of egg white which, once set by heat, stays set. The phenolics and ureas are important thermosetting plastics.

The other group contains the thermoplastics. These are soft when heated, hard when cooled, but can be softened and hardened repeatedly. You can compare them to sulfur and candle wax. The polyethylenes, polystyrenes, vinyls, and acrylics are in the thermoplastics "family."

THE MAKING OF A TYPICAL THERMOPLASTIC VINYL

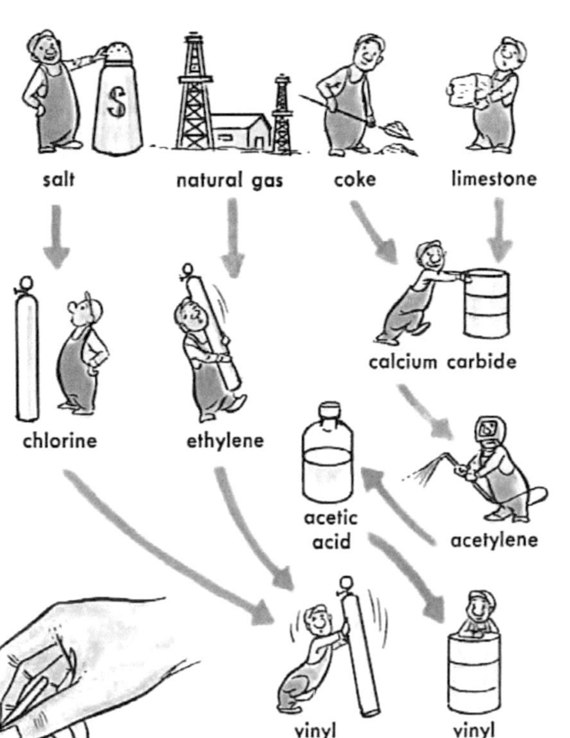

HEAT SHAPING. THERMOPLASTICS BECOME SOFT WHEN HEATED. YOU CAN THEN SHAPE THEM AT WILL.

BRING A POT OF WATER TO A BOIL. DROP IN AN OLD VINYL RECORD. WHEN SOFT, SHAPE IT WITH TWO LONG STICKS. IT BECOMES HARD AGAIN WHEN IT IS REMOVED FROM THE HOT WATER AND COOLED.

YOU CAN MAKE UNUSUAL DECORATIONS FOR THE WALLS OF YOUR GAME ROOM FROM OLD RECORDS.

MOLDING PLASTICS

1 CUT UP A SMALL AMOUNT OF SOFT PLASTIC. HEAT IT IN AN OLD TEASPOON.

2 SCRAPE SOFTENED PLASTIC ONTO A GLASS PLATE. PRESS A PENNY INTO IT. YOU GET A PERFECT MOLD.

Working out Chemical Equations

YOU HAVE DONE a great number of experiments by now. You have worked with gases, liquids, and solids. You have precipitated and decanted, filtered and distilled. As you think back over the experiments you will discover that they fall into four main groups of chemical reactions.

The simplest of these reactions is the DIRECT COMBINATION. In this, two or more substances combine to form a single more complex substance, as when iron and sulfur form iron sulfide:

$$Fe + S \rightarrow FeS$$

or when quicklime (calcium oxide) reacts with water to make slaked lime (calcium hydroxide):

$$CaO + H_2O \rightarrow Ca(OH)_2$$

DECOMPOSITION is the opposite of chemical combination. In this, a substance is broken down into simpler substances. This was the case when you separated the two elements found in water:

$$H_2O\ H_2O \rightarrow 2H_2\uparrow + O_2\uparrow$$

or when you made oxygen from hydrogen peroxide:

$$H_2O_2\ H_2O_2 \rightarrow 2H_2O + O_2\uparrow$$

In a SINGLE DISPLACEMENT, one element takes the place of another in a compound, as when you made hydrogen from zinc and hydrochloric acid:

$$Zn + HCl\ HCl \rightarrow H_2\uparrow + ZnCl_2$$

or when you set copper free by dropping a nail in a solution of copper sulfate:

$$Fe + CuSO_4 \rightarrow Cu\downarrow + FeSO_4$$

In a DOUBLE DISPLACEMENT, the two compounds change partners with each other. Think of the time when you precipitated silver chloride from solutions of salt and silver nitrate:

$$NaCl + AgNO_3 \rightarrow AgCl\downarrow + NaNO_3$$

or when you mixed Epsom salt and washing soda:

$$MgSO_4 + Na_2CO_3 \rightarrow MgCO_3\downarrow + Na_2SO_4$$

In studying the chemical shorthand above, you notice that, in every instance, there is an equal number of atoms of each element on either side of the arrow that indicates that a reaction takes place. Because of this equal arrangement, these chemical descriptions are called equations.

Many of these equations are scattered throughout this book. Many more are found in advanced chemistry textbooks. But very often, a chemist has to work out an equation from scratch.

Let's say you want to figure out the equation for dissolving aluminum foil in hydrochloric acid. You write out a trial equation:

$$Al + HCl \rightarrow AlCl + H\uparrow$$

But is AlCl correct? Look at the valence chart on page 75. Aluminum has three valence bonds, chlorine only one. One Al atom therefore takes on three Cl atoms, and aluminum chloride must be $AlCl_3$. H isn't right, either. Hydrogen exists in the free state only in molecules containing two atoms (H_2). So you change the equation to this:

$$Al + HCl \rightarrow AlCl_3 + H_2\uparrow$$

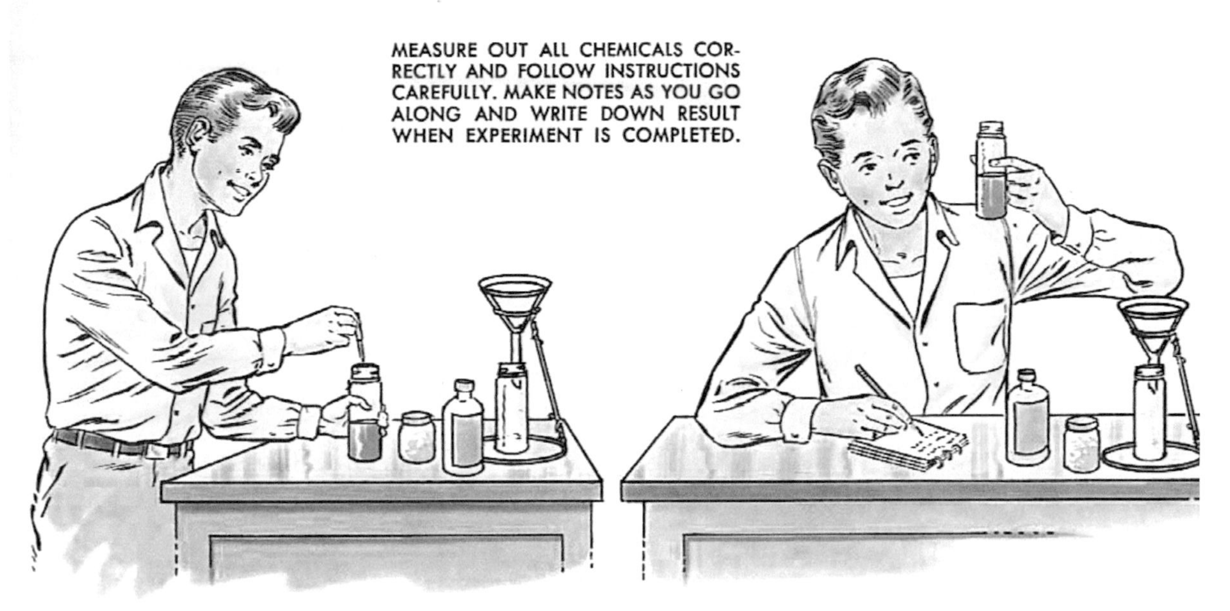

Now you need an amount of HCl that will give you Cl by the 3's and H by the 2's. 6HCl will do this. So you write in 6HCl and change the rest until the equation balances:

$$2Al + 6HCl \rightarrow 2AlCl_3 + 3H_2 \uparrow$$

Equations tell you what happens — but they tell far more than that.

Take the simple equation:

$$Fe + S \rightarrow FeS$$

APPROXIMATE ATOMIC WEIGHTS FOR CALCULATIONS					
Element	Symbol	Atomic Weight	Element	Symbol	Atomic Weight
ALUMINUM	Al	27	MAGNESIUM	Mg	24
BORON	B	11	MANGANESE	Mn	55
CALCIUM	Ca	40	NITROGEN	N	14
CARBON	C	12	OXYGEN	O	16
CHLORINE	Cl	36	POTASSIUM	K	39
COPPER	Cu	64	SILICON	Si	28
HYDROGEN	H	1	SILVER	Ag	108
IODINE	I	127	SODIUM	Na	23
IRON	Fe	56	SULFUR	S	32
LEAD	Pb	207	ZINC	Zn	65

This not only tells you that iron and sulfur make iron sulfide but also that it takes one iron atom and one sulfur atom to produce one molecule of FeS. Further, by inserting the atomic weights for the two elements from the chart on page 107, the equation tells you how much iron and sulfur are needed and how much iron sulfide you should get:

$$Fe + S \rightarrow FeS$$
$$56 \quad 32 \quad 56 + 32 = 88$$

You can use the atomic weight numerals to indicate numbers of grams or any other unit of weight. By dividing by 16 you get the number of grams you used for experiment on page 22.

Now take a more complicated equation.

Let's say you want to produce magnesium carbonate. The chart of solubilities on page 108 tells you that $MgCO_3$ is insoluble. You should therefore be able to precipitate it from a soluble magnesium salt — the sulfate, for instance — and soluble sodium carbonate:

$$MgSO_4 + Na_2CO_3 \rightarrow MgCO_3 \downarrow + Na_2SO_4$$

Now you need to know how much $MgSO_4$ and how much Na_2CO_3 you need, and how much $MgCO_3$ you will get.

Before you start figuring from the equation above, check the chart on page 108, top right. Here you will discover that each molecule of magnesium sulfate has seven molecules of water of hydration ($7H_2O$) attached to it, and each sodium carbonate molecule,

(CONTINUED ON PAGE 108)

SOLUBILITY OF SALTS AND HYDROXIDES

NITRATES—**SOLUBLE**—WITHOUT EXCEPTIONS.

ACETATES—**SOLUBLE**—WITHOUT EXCEPTIONS.

CHLORIDES—**SOLUBLE**—EXCEPT Ag, Hg (MERCUROUS), AND Pb.

SULFATES—**SOLUBLE**—EXCEPT Pb, Ba, Sr (Ca, Ag AND Hg SLIGHTLY SOLUBLE).

NORMAL CARBONATES, PHOSPHATES, SILICATES, SULFIDES—**INSOLUBLE**—EXCEPT Na, K, NH_4.

HYDROXIDES—**INSOLUBLE**—EXCEPT Na, K, NH_4, Ba. (Ca AND Sr SLIGHTLY SOLUBLE.)

WATER OF HYDRATION
(WATER OF CRYSTALLIZATION)

$AgNO_3$	$NH_4Al(SO_4)_2 \cdot 12H_2O$
$CaCl_2 \cdot 6H_2O$	NH_4Cl
$(CaSO_4)_2 \cdot H_2O$	$Na_2B_4O_7 \cdot 10H_2O$
$CuSO_4 \cdot 5H_2O$	$Na_2CO_3 \cdot 10H_2O$
$FeCl_3 \cdot 6H_2O$	$NaCl$
$FeCl_2 \cdot 4H_2O$	$NaHCO_3$
$FeSO_4 \cdot 7H_2O$	$NaHSO_4 \cdot H_2O$
$KAl(SO_4)_2 \cdot 12H_2O$	$NaOH$
KNO_3	$Na_2SO_4 \cdot 10H_2O$
$MgSO_4 \cdot 7H_2O$	$Na_2S_2O_3 \cdot 5H_2O$

Equations—Continued

ten molecules of water ($10H_2O$). These do not enter into the chemical reaction — but you have to include them in the weight of the chemicals.

Write the atomic weight below each element. Then figure the molecular weight of each compound by adding the atomic weights of all the atoms found in the molecule.

This is what you get:

Mg	S	O_4	$\bullet 7H_2O +$	Na_2	C	O_3	$\bullet 10H_2O \rightarrow$
24	32	16x4	7x18	23x2	12	16x3	10x18
24+	32+	64 +	126	46 +	12 +	48 +	180
	246					286	

Mg	C	$O_3\downarrow +$	Na_2	S	O_4	$17H_2O$
24	12	16x3	23x2	32	16x4	17x18
24+	12+	48	46 +	32+	64	306
84				142		306

(When a formula contains subscripts — the small numerals that indicate how many of a kind — be certain to multiply the atomic weight by the number indicated by the subscript. In cases where the formula is preceded by a large number, be sure to multiply the molecular weight by this number.)

Your finished calculation tells you that 246 g (or 24.6 g or 2.46 g) of magnesium sulfate crystals and 286 g (or 28.6 g or 2.86 g) of sodium carbonate crystals will give you 84 g (or 8.4 g or .84 g) magnesium carbonate.

When you get even deeper into chemical mathematics you will be able to figure out the percentage of elements in a compound for which you know the formula, or the formula of a compound when you know the percentage of elements, or the numbers of liters of a gas you prepare in a chemical reaction.

What's Ahead in Chemistry?

THE CHEMICAL WONDERS of today are amazing enough — but they are like nothing compared to the wonders the future holds in store for the welfare of all humanity.

FOOD — The fertilizers of tomorrow will greatly increase the crops grown on farms throughout the world. Insect and disease-destroying chemicals will make cattle and poultry healthier and better producers of meat, milk, and eggs. Chemicals unknown today will make it possible to keep food fresh without refrigeration in any climate.

HOMES — The houses of the future will be built of more durable materials than any we have today. Floors and wall covering will last almost indefinitely. New paints will add never-fading colors.

CLOTHING — Many more man-made fibers will be added to those we use today — fibers with longer wear; fabrics that are cool in summer, warm in winter, easy to keep clean.

HEALTH — The miracle drugs of today have wiped out diseases that ranked among our greatest killers just a few years ago. In years to come many more diseases will disappear from the surface of the world under the onslaught of still more effective drugs created in the chemical laboratory.

TRAVEL — Much of the travel of the future will be at supersonic speeds. Planes and rockets will require materials that can stand tremendous heat and new fuels capable of producing enormous energy. Chemistry will provide them.

ATOMIC ENERGY — The force hidden in the atom will be turned into light and heat and power for everyday uses. Chemists of the future, working with their brother-scientists, the physicists, will find new ways of harnessing and using the atoms of numerous elements — some of them unknown to the scientists of today.

Do you want to share in the making of that astonishing and promising future?

If you have enjoyed performing the experiments in this book, figuring out formulas and equations, jotting down observations, you are the kind of person who has the qualifications for making a successful career in chemistry.

If you care to look further into the matter, speak to your science teacher about it and drop a line to one or all three of the organizations mentioned below and ask for their pamphlets on becoming a chemist:

American Chemical Society,
1155 16th Street, N. W., Washington 6, D. C.

American Institute of Chemical Engineers,
25 West 45th Street, New York 36, N. Y.

Manufacturing Chemists' Association,
1625 I Street, N. W., Washington 6, D. C.

But whatever you decide for the future, keep up your interest in chemistry as a hobby. In addition to giving you fun and enjoyment, your chemical hobby will sharpen your powers of observation and reasoning and train your mind for whatever occupation you decide upon for a lifework.

CHEMISTRY AS A LIFE-WORK ENABLES YOU TO CONTRIBUTE TO THE WELFARE OF MANKIND.

Where to Get Chemicals and Equipment

A GREAT MANY of the experiments in this book can be performed with equipment found around the house: water glasses, custard cups, jars, bottles, cans, and funnel. For the rest, the following pieces of regular chemical laboratory equipment are needed:

- 6 test tubes, regular, 150 mm x 16 mm
- 3 test tubes, Pyrex, 150 mm x 16 mm
- 1 test tube brush, small
- 3 wide-mouth bottles, 4 ozs.
- 6 ft. glass tubing, 6 mm outside diameter
- 3 ft. rubber tubing, 3/16" inside diameter
- 2 No. 0 rubber stoppers, one hole
- 1 No. 5 rubber stopper, one hole
- 3 No. 5 rubber stoppers, two holes
- 1 triangular file, 4"
- 1 glass stirring rod, 5"
- 1 pkg. filter paper, 12.5 cm, 50 pieces
- 1 vial litmus paper strips, blue
- 1 vial litmus paper strips, red

If you can not secure this equipment locally, write to one of the companies below asking for price list or catalog, including cost of catalog where called for. When you receive the answer, mail your order and the correct amount by bank check or money order.

Science Mail Co., 17-33 Murray St., Whitestone 57, N. Y. (Price list free)
Winn Chemical Co., 124 West 23rd St., New York 11, N. Y. (Catalog 25¢)
N. Y. Scientific Supply Co., 28 West 30th St., New York 1, N. Y. (Catalog 50¢)
Home Lab Supply, 511 Homestead Ave., Mount Vernon, N. Y. (Price list free)
Biological Supply Co., 1176 Mt. Hope Ave., Rochester 20, N. Y. (Catalog 25¢)
A. C. Gilbert Co., P. O. Box 1610, New Haven 6, Conn. (Price list free)
Bio-Chemical Products, 30 Somerset St., Belmont, Mass. (Catalog 25¢)
Laboratory Sales, P. O. Box 161, Brighton, Mass. (Catalog 25¢)
The Porter Chemical Co., Hagerstown, Md. (Price list free)
Tracey Scientific Laboratories, P. O. Box 615, Evanston, Ill. (Catalog 25¢)
National Scientific Co., 13 South Park Ave., Lombard, Ill. (Catalog 35¢)
Hagenow Laboratories, Manitowoc, Wis. (Catalog 20¢)

IF YOU DECIDE TO USE REGULAR LABORATORY WARE IN YOUR HOME LAB, GET PRICE LIST FROM SUPPLIER.

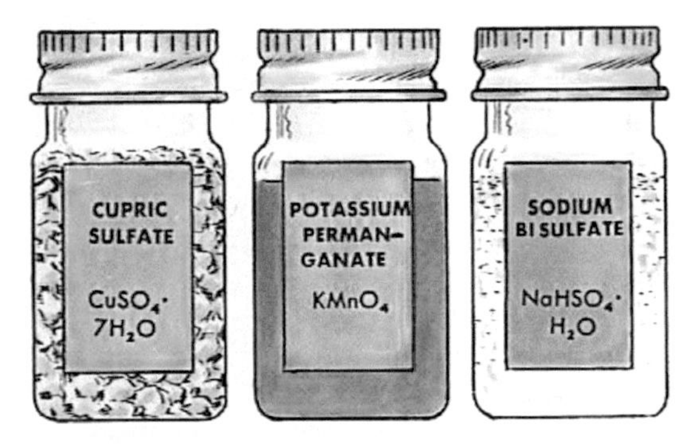

CHEMICALS FOR HOME EXPERIMENTS ARE AVAILABLE IN JARS OF UNIFORM SIZE, ALL PROPERLY LABELED.

WHENEVER YOU NEED a chemical for one of the experiments described in this book, check the list of common chemicals on page 111 to find out where to buy it.

All of these chemicals are, of course, available through chemical supply houses. The trouble is that many of these houses do not sell to individuals but only to schools and established laboratories. Also, the chemicals usually come in a standard amount of 1/4 lb. — or even 1 lb. — where, in home experiments, you would need 1 ounce or less. The same is often the case when you buy chemicals in a local store. The minimum-sized packages or jars may be so large that you couldn't possibly use up the contents in a year of experiments. You will probably also have to repack what you actually need into glass jars of suitable size for efficiency and to fit your storage space.

Because of this and the inconvenience of having to shop around, you may find it advantageous to buy your chemicals by the kit, in uniform-sized screw-top glass containers. Such kits are available in the science department of many hobby and model supply stores.

Chem-Kit No. 1 contains the ten chemicals marked ■ on the opposite page. Chem-Kit No. 2 contains the ten chemicals marked □. The kits contain sufficient amounts of chemicals to perform each experiment many times over.

You can also make up your own set of chemicals in amounts suitable for home experiments by getting them from one of the companies listed to the left. Be certain to add the cost of the catalog when you write for one and to send the correct amount when you order.

Common Chemicals and Their Formulas

	CHEMICAL NAME	FORMULA	COMMON NAME	WHERE TO BUY
□	ACETIC ACID	$CH_3COOH + H_2O$	5% solution: white vinegar	Grocery
■	AMMONIUM CHLORIDE	NH_4Cl	sal ammoniac	Drug store
□	AMMONIUM HYDROXIDE	$NH_4OH + H_2O$	10% solution: household ammonia 27% solution: strong ammonia	Grocery Drug store
□	BORIC ACID	H_3BO_3	boric acid	Drug store
■	CALCIUM CARBONATE	$CaCO_3$	chunks: marble, limestone powder: precipitated chalk	Builders' supplies Drug store
□	CALCIUM HYDROXIDE	$Ca(OH)_2$	slaked lime, garden lime	Hardware store
□	CALCIUM OXIDE	CaO	quicklime	Builders' supplies
□	CALCIUM SULFATE	$(CaSO_4) \cdot H_2O$ $CaSO_4 \cdot 2H_2O$	plaster of Paris gypsum	Hardware store Chemical supplies
□	CARBON TETRACHLORIDE	CCl_4	carbon tet	Hardware store
□	COPPER SULFATE	$CuSO_4 \cdot 5H_2O$	blue vitriol	Drug store
□	FERROUS SULFATE	$FeSO_4 \cdot 7H_2O$	iron sulfate, green vitriol, copperas	Drug store
□	GLUCOSE	$C_6H_{12}O_6 + H_2O$	solution: corn syrup	Grocery
□	HYDROCHLORIC ACID	$HCl + H_2O$	25% solution: muriatic acid	Hardware store
□	HYDROGEN PEROXIDE	$H_2O_2 + H_2O$	3% solution: peroxide	Drug store
■	IRON, METAL, POWDER	Fe	powdered iron	Chemical supplies
□	MAGNESIUM, METAL	Mg	magnesium ribbon	Chemical supplies
□	MAGNESIUM SULFATE	$MgSO_4 \cdot 7H_2O$	Epsom salts	Drug store
■	MANGANESE DIOXIDE	MnO_2	pyrolusite	Hardware store (flashlight battery)
□	NAPHTHALENE	$C_{10}H_8$	moth balls	Hardware store
■	PHENOLPHTHALEIN	$C_6H_4COOC(C_6H_4OH)_2$	phenolphthalein	Drug store
□	POTASSIUM ALUMINUM SULFATE	$KAl(SO_4)_2 \cdot 12H_2O$	alum, potassium alum	Drug store
□	POTASSIUM FERROCYANIDE	$K_4Fe(CN)_6 \cdot 3H_2O$	potassium ferrocyanide	Chemical supplies
■	POTASSIUM IODIDE	KI	potassium iodide	Drug store
□	POTASSIUM NITRATE	KNO_3	saltpeter, niter	Drug store
□	POTASSIUM PERMANGANATE	$KMnO_4$	potassium permanganate	Drug store
□	SALICYLIC ACID	$C_6H_4OHCOOH$	salicylic acid	Drug store
□	SILVER NITRATE	$AgNO_3$	lunar caustic	Drug store
□	SODIUM BICARBONATE	$NaHCO_3$	baking soda, bicarb	Grocery
■	SODIUM BISULFATE	$NaHSO_4 \cdot H_2O$	82% of Sani-Flush®	Grocery
□	SODIUM CARBONATE	$Na_2CO_3 \cdot 10H_2O$ $Na_2CO_3 \cdot H_2O$	sal soda, crystal washing soda concentrated washing soda	Grocery Grocery
□	SODIUM CHLORIDE	$NaCl$	salt, table salt	Grocery
□	SODIUM HYDROXIDE	$NaOH$	lye, caustic soda, Drano®	Grocery
□	SODIUM HYPOCHLORITE	$NaClO + H_2O$	5% solution: laundry bleach, Clorox®	Grocery
□	SODIUM POTASSIUM TARTRATE	$NaKC_4H_4O_6 \cdot 4H_2O$	Rochelle salt	Drug store
□	SODIUM SILICATE	$Na_2SiO_3 + H_2O$	solution: water glass	Hardware store
□	SODIUM TETRABORATE	$Na_2B_4O_7 \cdot 10H_2O$	borax	Drug store
■	SODIUM THIOSULFATE	$Na_2S_2O_3 \cdot 5H_2O$	hypo	Photo store
□	SUCROSE	$C_{12}H_{22}O_{11}$	cane sugar	Grocery
■	SULFUR	S	powder: flowers of sulfur block: sulfur candle	Drug store Hardware store
■	ZINC, METAL	Zn	zinc	Hardware store (flashlight battery)
□	ZINC CHLORIDE	$ZnCl_2 + H_2O$	tinners' fluid	Hardware store

Note: Chemicals marked □—many of them liquids—are most easily secured in local stores. Chemicals marked ■ are found in Chem-Kit No. 1, chemicals marked □ in Chem-Kit No. 2 (see opposite page).

Index

Made in United States
Troutdale, OR
12/17/2024